Integrating Food Science and Engineering Knowledge Into the Food Chain

The ISEKI-Food book series is a collection where various aspects of food safety and environmental issues are introduced and reviewed by scientists specializing in the field. In all of the books a special emphasis is placed on including case studies applicable to each specific topic. The books are intended for graduate students and senior level undergraduate students as well as professionals and researchers interested in food safety and environmental issues applicable to food safety.

"ISEKI-Food" is an acronym for "Integrating Safety and Environmental Knowledge Into Food Studies". Born from an Erasmus network that ran until 2014, the ISEKI-Food Association sustains the network activities, continuing to promote the enlargement of this book series with up-to-date topics. The association has members from more than 60 countries from all continents and its main objectives are:

Establishment and maintenance of a network among universities, research institutions and companies in the food chain by

- Promoting synergies between research, education and industry
- Development of a virtual community of experts in the field of food, with communication to the general public
- Establishment of a framework of agreements among partners, fostering the mobility of students and staff
- Stimulating the development of further, related projects

Working towards the quality assurance of food studies by

- Fine tuning curricula
- Developing teaching materials and teaching methods
- Cooperation in the implementation of quality criteria in the food chain
- Accreditation of food studies

Saverio Mannino

Guidelines for Laboratory Quality Managers

Springer

Saverio Mannino
Formerly, Department of Food Science
Technology and Microbiology
University of Milan
MILANO, Italy

ISSN 2512-2223 ISSN 2512-2258 (electronic)
Integrating Food Science and Engineering Knowledge Into the Food Chain
ISBN 978-3-031-11723-7 ISBN 978-3-031-11724-4 (eBook)
https://doi.org/10.1007/978-3-031-11724-4

This Springer imprint is published by the registered company Springer Nature Switzerland AG
The registered company address is: Gewerbestrasse 11, 6330 Cham, Switzerland

This book is dedicated to my granddaughters Michelle and Sofie.

Preface

This book attempts to bring together salient and topical aspects of laboratory operations to help mangers to develop a quality system for managing laboratory samples from the time of collection to the final storage or disposal, in a manner that will ensure reliable results. The author hopes that the book may help the reader to obtain a clear overview of the complex science involved and its application to the production of reliable results, as required by ISO 17025:2017.

The laboratory operations are considered analogous to a manufacturing process where the industrial process is substituted by the analytical process whose production are analytical results. Process control is then an essential element of the quality management system. It refers to control of the activities employed in the handling of samples and the examination processes in order to monitor the accuracy and precision of the complete analytic process. The goal of quality control is to detect, evaluate, and correct errors due to test system failure, environmental conditions, or operator performance, before results are reported. This book provides practical tools for managing more effectively an analytical laboratory, and guides the reader to carefully evaluate the variability of the analytical process in order to improve it. The guiding principle is that quality results are not produced but the quality of results is improved by controlling all stages of the process including sampling.

Chapter 4 guides the user through the required statistics to solve many problems related to laboratory activity and to carefully evaluate the variability in the process in order to improve it. Statistical control and process capability are treated in Chap. 5. Finally, risk analysis is covered in Chap. 7. Whenever possible, chapters have been designed to stand alone.

Furthermore, it is assumed that readers will rely upon commercial software to do statistical analysis.

The book is recommended to students following laboratory courses, teachers, laboratory managers, technicians, and anyone engaged in the laboratory accreditation process.

MILANO, Italy Saverio Mannino

Acknowledgments

Many university staffers have contributed to the preparation of this book that is part of lectures prepared for the course of food analysis. I am appreciative of those who helped to define major topics for this book and develop a structure in which to present them. They were Maria Stella Cosio, Matteo Scampicchio, Susanna Buratti, and Simona Benedetti, and they also encouraged me to write this book and gave me invaluable assistance throughout the process.

Of course, none of them are responsible for errors and omissions since the final decision on content is mine.

Contents

Chapter 1
Introduction and ISO17025:2017

Chemical measurements are playing a crucial role in modern society since is the basis of important decisions. In fact, based on such decisions, national and international regulations concerning commercial, legal, medical and environmental impact are set in force. For what concern foodstuff, their value in trade depend strictly on the measurements that determine the degree of quality. However, results of measurements must be reliable and to be trustful should be obtained in competent and well-organized laboratories working with quality system in place.

The ISO/IEC 17025:2017 "General requirements for the competence of testing and calibration laboratories" is the global quality standard for testing and calibration laboratories.

This standard specifies the performance requirements for the procedures in a quality system and laboratories can determine their own pattern for carrying out procedures.

Laboratories that, at assessment, can demonstrate compliance with ISO/IEC 17025 must prove to operate using sound management practices and that are technically competent to perform specific tests, and/or measurements as well as at the same time are able to generate technically valid results for which they hold accreditation.

The last version of the standard of the year 2017 specifies the general requirements for the competence, impartiality, and consistent operation of testing and calibration laboratories and it is applicable to all organizations performing testing, calibration, and/or sampling. With respect to the old one it provides an update of terminology and scope, an alignment with the recent standard ISO 9001, and the introduction of the concept of risk-based thinking.The ISO/IEC 17025:2017 is the basis for the accreditation from an accredited body. Accreditation involves a third party or an accreditation body attesting the technical competence of a laboratory based on a set of accepted standards. Accreditation bodies that recognize the competence of testing and calibration laboratories use ISO/IEC 17021-1:2005 as the basis for their accreditation.

S. Mannino, *Guidelines for Laboratory Quality Managers*, Integrating Food Science and Engineering Knowledge Into the Food Chain 14,
https://doi.org/10.1007/978-3-031-11724-4_1

The ISO/IEC 17025:2017 is divided into eight clauses, two annexes A (Metrological traceability) and B (Management System Options), and one bibliography section. The most important clauses are clauses 6,7 and 8, describing management and technical requirements. In addition to official requirements, these clauses also include notes with further explanations and recommendations.

Below are reported some brief considerations on each point of the norm ISO17025:2017 and where appropriate some difference with the old norm ISO 17025:2005 are discussed.

1.1 Clause 1: Scope

The standard specifies the general requirements for the competence, impartiality and consistent operation of laboratories and covers the technical activities as well as the management and organizational aspects of laboratories.

1.2 Clause 2: Normative References

The ISO/IEC Guide 99, International vocabulary of metrology – Basic and general concepts and associated terms and the ISO/IEC 1700, Conformity assessment – Vocabulary and general principles – are listed as references for vocabulary and conformity assessment.

1.3 Clause 3: Terms and Definitions

Terminology has been updated and includes the changes that have been made in the International Vocabulary of Metrology (VIM). The term "laboratory" has been added: it refers to the bodies that perform one of the following activities such as testing, calibration, and/or sampling, associated with subsequent testing or calibration. Other terms and definitions concern among others: impartiality, complaints, inter and intra-laboratory comparison, proficiency testing, decision rule, verification and validation. In the glossary selected important terms from ISO5127:17 – Information and Documentation – Foundation and Vocabulary – are reported.

1.4 Clause 4: General Requirements

Most of the requirements are similar to those specified in the ISO Standard 9001:2000 and include impartiality and safeguard of confidentiality.

1.4.1 *Impartiality*

Presence of objectivity and no conflicts of interest that can adversely influence the laboratory activities should exist. If a potential risk to impartiality is identified, the laboratory shall prove how can eliminate or minimize such risk. Approaches and methods should be documented.

In other words, the organizational structure should be such that departments having conflicting interests do not adversely influence the laboratory's work quality. All personnel should be free from any commercial or financial pressure that could adversely impact the quality of calibration and test results.

1.4.2 *Confidentiality*

Any information obtained or created during the normal performance of laboratory activities, must be protected by confidentiality. It includes data and information generated in any format: electronic storage and/or transmission.

1.5 Clause 5.0: Structural Requirements

Most of the requirements come from the ISO/IEC Guide 25:1990. This section ensures that the roles and responsibilities of the laboratory, the management, and key personnel are defined. The laboratory shall be a legal entity that define and documents the range of the laboratory activities. In addition, it defines: the organization and management structure of the laboratory; responsibility, authority and interrelationship of all personnel; documentation of its activities.

Other key points are:

- Identification of management that has overall responsibility for the laboratory.
- Presence of personnel that have the authority and resources to implement, maintain and improve the management system.

1.6 Clause 6.0: Resources Requirements

1.6.1 *General*

This section highlights the importance of availability of "personnel, facilities, equipment, systems and support services necessary to manage and perform its laboratory activities.

Key points are:

1.6.2 Personnel

All personnel of the laboratory should be sufficiently competent for the work to be done. There should be procedures to identify training needs of the personnel and job descriptions or similar regulations to describe the competence and duties of the personnel, including impartiality. The relation between management and subordinate staff should be clearly stated. Records of the qualification of personnel must be kept up to date. This section is correlated to section 4.1 and 5. See the tip HUMAN RESOURCES in the next chapter for more details.

1.6.3 Facilities and Environmental Conditions

– The laboratory must keep accommodation and environmental conditions under control. Adequate monitoring is part of this. Procedures will describe the activities and regulations taken.
– In the risk assessment for each method particular attention should be paid to the suitability of facilities and environmental condition and to their control.
– Control to access to lab areas and prevention of contamination must be maintained

1.6.4 Equipment

– The laboratory is required to have procedures for the receipt, handling, protection etc. of testing and calibration equipment. Procedures must describe activities and regulations taken. The laboratory should also verify the correct performance of all the equipment and their influence on the results. In the note 1, the use of reference materials and certified reference materials is highly recommended. As guidance for this aspects the ISO 17034, the ISO Guide 33 and ISO Guide 80 should be consulted.
– Records of all instrumentation should be retained
– It is required that the laboratory be furnished with all items needed for sampling and testing. There is adequate maintenance for all items and there are procedures to cover all relevant aspects of dealing with them. Records will be accessible for all relevant aspects and working with the equipment will be in accordance with the procedures.
– Personnel should receive sufficient training.

1.6.5 Metrological Traceability

Traceability can be defined as an unbroken record of documentation ("documentation traceability") or an unbroken chain of measurements and associated uncertainties ("metrological traceability"). All test and calibration equipment will be verified or calibrated at regular intervals. It is assumed that these are planned activities of the laboratory and has procedures for dealing with reference materials and standards. All measurements should be traceable to the International System od Units (SI).

The International Vocabulary (VIM) defines traceability as "Property of a measurement result whereby the result can be related to a reference through a documented unbroken chain of calibrations, each contributing to the measurement uncertainty".

Thus, calibration and the use of certified reference material are the central tools in establishing traceability.

Traceability, as already said, refers to assurance that the measurement results can be related to a reference through a documented, unbroken chain of comparisons. For example, the bias, precision and accuracy of testing can be determined by testing a certified reference material and comparing the laboratory results with the certified value. The certified value of the reference material is generally reported with uncertainty such that the comparison is of statistical significance.

As example we can use the thermometer. To establish and maintain traceability, the readings of a thermometer can be compared to a fixed-point temperature (e.g., ice-melting point) or a reference thermometer at a fixed temperature – this testing process is often called verification, performance validation or calibration.

Once a thermometer's accuracy is authenticated, it can serve as a reference to establish traceability for other thermometers. This process can be continued, providing an unbroken chain of measurements from the final thermometer all the way back to the international standards.

Furthermore, traceability allows the comparability and the harmonization of results between analytical laboratories. Hence, demonstration of traceability is the primary objective in chemical measurements.

1.6.6 Externally Provided Products and Services

The laboratory shall ensure that only suitable externally provided products and services that affect laboratory activities.

Services and supplies delivered by third parties should not adversely influence the quality and effectiveness of laboratory operations.

Key Points

- The competence of the subcontracted party should be guaranteed, through a documented quality system.
- The subcontracting laboratory is responsible to the customer for the subcontractor's work.
- Suppliers should be selected and formally evaluated to guarantee that services and supplies are of adequate quality.

- Records of the selection and evaluation process should be maintained.
- The quality of incoming material should be verified against predefined specifications.
- There must be procedures in the laboratory, which ensure that companies meeting predefined standards supply, quality-relevant products and services.

This requirement is connected with clause 8.5: Actions to address risk and opportunities.

1.7 Clause 7.0: Process Requirements

This section is the longest of the norm and contains eleven sub-clauses deployed as follows:

7.1 Review of requests, tenders, and contracts
7.2 Selection, verification, and validation of methods
7.3 Sampling
7.4 Handling of test or calibration items
7.5 Technical records
7.6 Evaluation of measurement uncertainty
7.7 Ensuring the validity of results
7.8 Reporting results
7.9 Complaints
7.10 Non-conforming work
7.11 Control of data-information management

Most of these clauses are included in scheme of the process of Annex II of the norm, reported below.

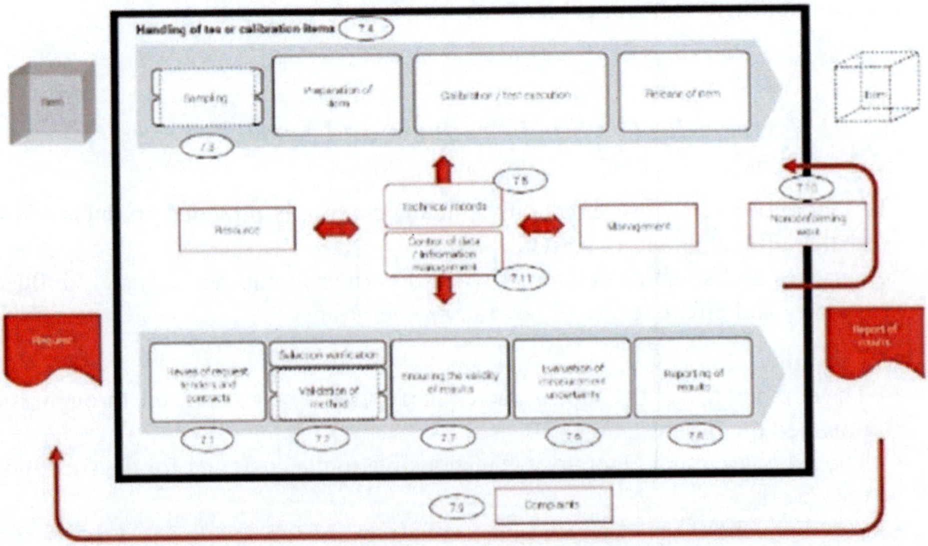

1.7.1 *Review of Requests, Tenders and Contracts*

This section describes how to ensure that requirements of requests, tenders and contracts are well defined, reviewed, understood, and documented.

Key Points

- The laboratory will run a documented system for dealing with requests, tenders and contract reviews.
- The laboratory manager's review should guarantee that the laboratory has the technical capability and resources to meet the requirements.
- The laboratory should select appropriate methods that meet customer's requirements
- Communication and cooperation with customers must be loyal and effective
- Changes in a contract should follow the same process as the initial contract.
- When a customer requires a declaration of result as a pass or fail, in or out of tolerance, a decision rule must be selected and agreed with the customer unless the decision rule is built-in or defined in the standard or specification. See clauses 7.8.3 and 7.8.6 for more details.

1.7.2 *Selection, Verification, and Validation of the Methods*

Laboratories shall use appropriate measurement procedures to meet customer requirements and apply appropriate methods and procedures in performing testing and calibration work. Laboratories must also have instructions for running equipment. International or other recognised specifications do not necessarily need to be supplemented or rewritten if the laboratory staff can apply them. In the event of new testing or calibration methods developed by the laboratory, it is required that the laboratory implement a validation study. Furthermore, it is required that the laboratory be able to estimate the uncertainty of its measurements. It is mandatory for the laboratory to control calculations and electronic data. Computer software especially that developed by the laboratory, is documented in sufficient detail and has proved to be suitable for use.

For more information on how to select, verify and validate a method see Chap. 3 of this book.

1.7.3 *Sampling*

Sampling is a defined procedure whereby a part of a substance, material or product is taken to provide for testing or calibration of a representative sample of the whole. Sampling may also be required by the appropriate specification for which the substance, material or product is to be tested or calibrated.

However, the laboratory shall have a sampling plan and procedures for sampling when it carries out sampling of substances, materials or products for subsequent testing or calibration. The sampling plan as well as the sampling procedure shall be available at the location where sampling is undertaken. Sampling plans shall be based, whenever reasonable, on appropriate statistical methods by using a risk based approach.

1.7.4 Handling of Test or Calibration Items

After sampling a series of procedures are required for handling the test and calibration items. Laboratory must have instructions to protect the integrity of the sample including handling till the subsequent analysis and storage or conditioning. Records of all operations must be maintained.

1.7.5 Technical Records

Technical records are important documents that contain useful information about the laboratory activities. Usually they contain data, results and identity of the personnel responsible of the activity. All records should be traceable. The clause *8.4 Control of records*, the general requirements for all records, also applies to this section.

1.7.6 Evaluation of Measurements Uncertainty

Every measurement is inexact and therefore requires a statement of uncertainty to quantify that inexactness. Laboratory shall identify all the contributions that have significance on the measurement uncertainty. For more information, the norm recommends the ISO IEC Guide 983, ISO 21748 and the ISO 5725 series. Chapter 4 reports an example of uncertainty measurement.

1.7.7 Ensuring the Validity of Results

The laboratory will run suitable programmes for internal and external quality control schemes. These will include statistical techniques, participation in proficiency testing programmes, use of standard reference material, quality control material, alternative instrumentation, intermediate check, replicates, blind samples and so on.

A proficiency test (PT) is simply, a method that you may use to validate a particular measurement process. It is an inter-laboratory comparison, in which a number of labs conduct testing methods within their own lab on the same material and report the results to the organizing party. Each individual laboratory is then evaluated for performance based on statistical calculations.

Results from proficiency testing are an indication of a laboratory's competence and are an integral part of the assessment and accreditation process. Proficiency testing programs may take many forms and standards for satisfactory performance vary depending on the field. Applicants and accredited laboratories are required to participate in relevant and available proficiency testing provided by organizations administering acceptable proficiency testing programs.

NOTE: ISO/IEC 17043 contains additional information on proficiency tests and proficiency testing providers. Proficiency testing providers that meet the requirements of ISO/IEC 17043 are considered to be competent.

1.7.8 Reporting Results

The requirements of test report mainly follow those of the old EN 45001 norm. An important change is, however, that opinions and interpretations may now be part of the test report if they are clearly separated from the test results. This is the second longest section of the norm. and present the following subsections: General (7.8.1), Common requirements for reports (test, calibration or sampling) (7.8.2), Specific requirements for test reports (7.8.3), Specific requirement for calibration certificates (7.8.4), Reporting sampling – specific requirements (7.8.5), Reporting statement of conformity (7.8.6), Reporting opinions and interpretation (7.8.7), Amendments to reports (7.8.8).

1.7.9 Complaints

This section describes how to ensure that any customer complaints are documented, evaluated, and adequately followed up. It is correlated with the clauses: 5 – Actions to address risks and opportunities and addressing nonconforming work, 7.10 Nonconforming work and Corrective actions.

Key Points

- There should be a policy and procedure for the resolution of complaints received from customers.
- Records of complaints and all steps taken when resolving the complaint should be maintained.

This includes documentation of investigations and corrective actions. The laboratory will have procedures for dealing with complaints. The feedback from their client should be used for continuous improvement of their services.

The laboratory should communicate with customers to clarify requests and get customer input. The laboratory should have a formal program to collect feedback from customers on an ongoing basis.

The laboratory should allow customers to audit the laboratory

- Procedures in the laboratory will regulate co-operation with its clients.

1.7.10 Nonconforming Work

This section requires laboratory to introduce general procedures for dealing with nonconforming services. Tests, calibrations, and other laboratory operations should conform to previously defined specifications such as laboratory specifications or client-defined specifications.

This section describes how to ensure that nonconforming test and/or calibration results are adequately followed up, and that actions are based upon the risk level established by the laboratory. It is correlated to point 8.5 "Actions to address risks and opportunities" and 9.7 "Corrective actions".

Key Points

- There should be a policy and process that come into effect when results are not conform to established procedures.
- Corrective actions should be taken immediately to avoid recurrence.
- The significance of nonconforming work should be evaluated, for example, the possible impact on other testing or calibration work.
- If necessary, customers should be notified.

1.7.11 Control of Data – Information Management

Laboratory data-information management include data and information in both computerized and no-computerized system. Data and information deal with systems, processes, organizations, services or products, and persons in the laboratory.

The data – information management system is correlated with the clauses: 8.2 Management system documentation and 8.3 Control of management system documents, 8.4 Control of records and 7.5 Technical records.

Key Points

The laboratory shall establish, implement and maintain a quality management system which is appropriate to its scope of activities. This quality management system will be documented in manual, procedures, calibration instructions, job descriptions and similar documents.

- The quality policy statement must be relevant and consistent with the scope of activities. There should be policies, standard procedures and work instructions to ensure the quality of test results.
- There can be a quality manual with policy statements that are issued and communicated by top-level management.
- Operate with a validated LIMS and protect the system from unauthorized access
- Check the data and calculation transfer

1.8 Clause 8.0: Management System Requirements

This section states that a laboratory has two options when implementing a management system: Option A and Option B.

Option A: This option lists the main requirements for implementing a Laboratory Management System (LMS) and is applicable for laboratory that do not have a management system in compliance with ISO9001. This means that the laboratory is free to implement a management system based on the requirements of ISO17025. In addition, the laboratory can choose to incorporate the requirements of ISO 9001 that are relevant for performing laboratory activities.

Option B: In the option B laboratory has to operate in accordance with ISO 9001, in a way that fulfills requirements 4–7 of the ISO 17025:2017 standard.

However, as it is stated in the norm, both options are intended to achieve the same results in the performance of the management system and compliance with clause 4–7.

Beside the two options, the norm presents the following sub clauses for Option A.

1.8.1 Management System Documentation (Option A)

This section describes a system of document control addressing issues like document approval and issuing as well as changes to documents. Of course, documents may be written on paper or they may be in electronic form.

Key Points

All documents related to the management system are uniquely identified and created, approved, issued, and changed following documented procedures. All official documents should be authorized and controlled.

Documents should be regularly reviewed and updated if necessary. The review frequency depends on the document itself. Typical review cycles are between one and three years. Changes to documents should follow the same review process as for the development of initial documents.

The ISO/IEC 17025 requires different types of documentation, as illustrated in the following documentation pyramid.

The Quality Manual is the top tier of the document hierarchy. It describes the approaches to achieve quality data. It also includes policy statements describing the laboratory's intention to conform to ISO/IEC 17025 requirements. However, laboratory manual it is not mandatory. A process or generic procedure describes how various quality requirements can be achieved. Standard operating procedures (SOPs) or Working Procedures are step-by-step instructions for how to exactly perform a specific task, such as calibrating a specific instrument.

Records are generated on a day-by-day basis, such as analytical results from product tests or calibration records of a balance.

1.8.2 Control of Management System Documents (Option A)

All documents should be properly controlled. For example, each change should be authorized and logged, and the updated document should get a new revision number or code.

Laboratory shall control all internal and external documents concerning the quality system. Documents in hard copy or digital can be procedures, specification, text book drawings, plan, etc. However, all documents should be uniquely identified, updated, and periodically reviewed. Documents include both internal, such as SOPs, quality manuals, and training plans and external documents, such as regulations, standards, test methods, and instrument operating manuals.

Key Points

- The laboratory will run a system for identification, collection, indexing, distribution and storage of records. Special requirements are defined for technical records, which are accumulations of data, and for other information on tests and calibrations, which have been performed.

- There should be procedures for identification, collection, indexing, storage, retrieval, and disposal of records
- Records should be stored such that their security, confidentiality, quality and integrity are ensured throughout the required retention time
- For technical records such as test reports of analytical measurements, original observations should be retained, along with processing parameters that will allow tracking final results back to the original observations.
- Record format can be hard copies or electronic media. There should be procedures to protect and back-up electronic records and to prevent unauthorized access.
- Records can be corrected if there are mistakes. The original record should be crossed out, but still visible.
- When electronic record systems are used, the same principle applies. The laboratory should ensure that original records are not overwritten by the system and that corrections are recorded together with the original records. Using a system that prevents overwriting original records and stores changes in an electronic audit trail that can be viewed and printed is highly recommended.
- Records to demonstrate conformity with ISO/IEC 17025 and as required by customers should be retained for a specific amount of time

Checklists, forms, templates, and examples help implement quality work effectively and consistently.

Official documents are created or acquired, reviewed, and approved prior to use.

Documents are uniquely identified with document and revision number, date of revision, and issuing authority.

- A quality list with all controlled documents is maintained by QA. The list includes document and revision number, title, date of issue, date of last review, and locations.
- Internal documents include page numbers and total number of pages on each page. Users of the documents are adequately trained before the documents are released.
- Current authorized versions of documents are readily available at the user's workspace.

1.8.3 Control of Records (Option A)

This section describes how to ensure that all records in a laboratory are uniquely identified, readily available when needed, and protected against unauthorized access for viewing or changing.

Development and maintenance of documentation should be controlled through document control and management procedures that are part of the management system.

1.8.4 Actions to Address Risks and Opportunities (Option A)

The laboratory shall consider risks and opportunities that could potentially directly harm the organization, reputation, staff members and so on. Actions should be taken to prevent or reduce impacts on the laboratory activities. No formal risk management system is required but laboratory should select appropriate methodology to address risk and opportunities. This section is correlated with the next one 8.6 Improvement.

1.8.5 Improvement (Option A)

This section describes how to ensure that the effectiveness of the management system is continually improved.

Key Points
- Areas of improvement become opportunities. Laboratory shall identify such areas through the review of the operational procedures, objectives, audit results, proficiency testing, customer satisfaction surveys and so on.
- Suggestions for improvements can came from audit reports, analysis of data, customer complaints and suggestions, corrective and preventive actions, and management reviews.
- Suggestions should be collected over time and reviewed by management for suitable actions.

1.8.6 Corrective Actions (Option A)

This section describes how to ensure that the root cause of nonconforming work or deviations from laboratory and management procedures are identified and that adequate corrective actions are selected, implemented, documented, and monitored. Procedures for corrective action will include cause analysis as well as standard paths.

Key Points
- Corrective actions can be triggered through nonconforming tests or other work, customer complaints, internal or external audits, management reviews, and observations by staff.
- Corrective actions should be selected and implemented to eliminate the specific problem and prevent recurrence of the same problem.
- As the first step in the process, the root cause of the nonconformity should be identified.
- The effectiveness of the corrective action should be monitored and evaluated.

- There should be a procedure to identify potential sources of nonconformities and define preventive actions to prevent occurrence of these nonconformities. Nonconformities may be technical or related to the management system. The objective is to reduce their likelihood of occurrence
- Procedures will be maintained which focus on preventive action in all relevant areas of activity of the laboratory.

1.8.7 Internal Audits (Option A)

Internal audits should verify that the laboratory complies with the norm and with internal technical and quality procedures. Internal audits are also an excellent preparation for external assessments and can help to continually improve the quality system. The laboratory will plan and conduce internal audits on a regular basis. These audits are required to address all elements of the quality management system. The results of the audits will be clearly reported and corrective action will be made within z an acceptable period. A risk based approach is required.

The section is correlated to clause 8. Management review, 8.5 – Addressing risk and opportunities and 8.6 Improvements. ISO 19011 gives guidance for internal audit.

Key Points
The laboratory should have a procedure and a schedule for internal audits. Internal audits can either cover the whole laboratory and all elements of the quality system at one specific period of time or can be divided into several subsections.

The schedule should be such that each element of the quality system and each section of the laboratory are audited yearly.

The audit program should be managed by the quality manager.

Audit follow-up activities should include corrective and preventive action plans (CAPA).

The effectiveness of the plans should be monitored.

1.8.8 Management Reviews (Option A)

Requirements in this section describe how to ensure the continued suitability and effectiveness of the quality system, policies, and testing and calibration procedures. The management of the laboratory is required to conduct management reviews of all elements of the quality management system. The results of audits, incoming complaints and other sources of information as well as corrective action taken and other issues of relevance will be referred to in the management review.

Key Points

- There should be a schedule and procedure for periodic management reviews. Recommended review frequency is once a year.
- The management review should include a discussion about the outcome of recent internal audits and external assessments, corrective and preventive actions, results of proficiency testing, customer complaints and feedback, and any recommendations for improvements.
- Management should decide on follow-up activities. These activities should be monitored for effectiveness.

Chapter 2
Essentials for Quality Management in a Chemical Testing Laboratory

In the previous chapter we saw that ISO17025:2017 requires the establishment in the laboratory of a quality management system. This chapter is concerned with the planning and organizing that must be completed to establish a good quality assurance program in an existing laboratory, which may already have some of the elements, but not all. In fact, many laboratories have various aspects of quality control in place but are deficient in quality assurance. Quality control is generally nothing but good scientific practice, but the elements of documentation, sample control, traceability, and so on, are often lacking in many laboratories.

However, a fundamental premise to implement a quality assurance program in a laboratory is to consider measurements as a process, analogous to a manufacturing process (Fig. 2.1). In effect, laboratories can be considered as services where the industrial process has been substituted by the analytical process.

It is then possible to apply the same principles of quality management system in the industries and the tools provided by the DMAIC (Define, Measure, Analyze, Improve, Control) and PDCA (Plan, Do, Check, Act) models.

The process may be brought into a state of statistical control, i.e., the individual measurements are defined by a statistical distribution and its characteristic precision and accuracy can be assigned to the data output. Obviously, quality control relates to all that is done to attain and maintain the state of statistical control.

2.1 Managing the Quality of Laboratory Testing Processes

There is extensive literature about quality management in testing laboratories. Topics frequently include Quality Control, Quality Assurance, Quality Assessment, Quality Improvement, and Quality Planning, which are all part of Quality

S. Mannino, *Guidelines for Laboratory Quality Managers*, Integrating Food Science and Engineering Knowledge Into the Food Chain 14,
https://doi.org/10.1007/978-3-031-11724-4_2

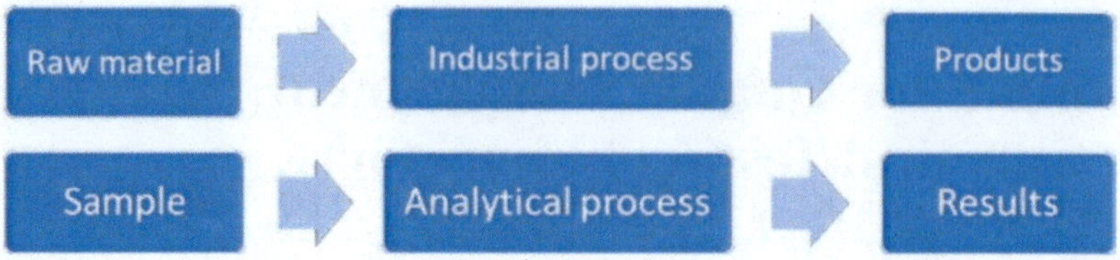

Fig. 2.1 Industrial and laboratory processes

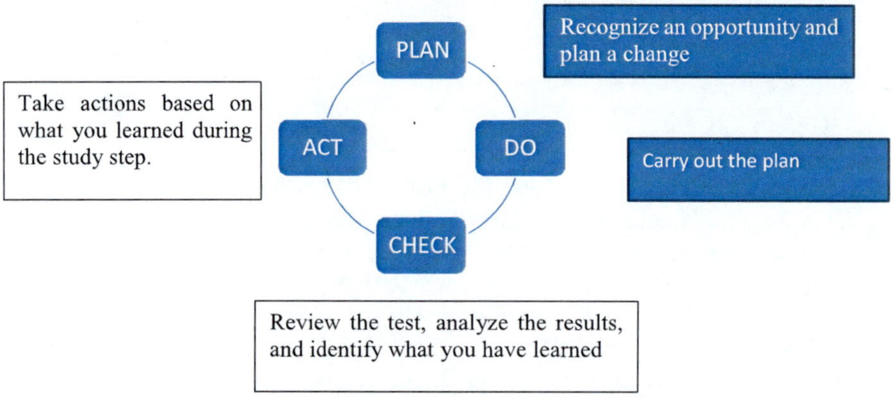

Fig. 2.2 The PDCA cycle

Management today. For laboratory scientists, it is a challenge to integrate all these programs, guidelines, standards, and tools into a cohesive Quality Management.

The ISO 17025:2017 standard adapts industrial principles and concepts specifically for application in testing laboratory, creating a global standard for quality and competence in testing laboratories.

2.2 Developing a Quality Management System (QMS)

The fundamental model for a quality management system is the Deming's Plan-Do-Check-Act cycle, which embodies the principles of scientific investigation and objective decision- making. W. Edwards Deming, who is considered by many to be the father of modern quality control, made PDCA very popular. The PDCA cycle is commonly presented as shown in Fig. 2.2.

2.2.1 PLAN

Planning helps to establish the objectives and processes necessary to deliver results in accordance with the expected output (the target or goals).

Careful planning keeps the quality team focused on the central issue it's working on.

2.2.2 *DO*

Implement the plan, execute the process, and make the product. If required, change the process to eliminate the defect. Be sure to have a back out plan in case the change does more harm than good.

Collect data for charting and analysis in the following "CHECK" and "ACT" steps.

2.2.3 *CHECK*

Monitor the changed process by using the same types of measurements that was used to identify the problem. Measurable quality improvement objectives make demonstrating process improvement easy. Charting data can make this much easier to see trends over several PDCA cycles and in order to convert the collected data into information. Information is what is needed for the next step "ACT".

2.2.4 *ACT*

Ensure the change is firmly established into the process and that the problem is fixed for good. Go back to the problem statement and confirm that the change eliminated the problem. In many cases, the fix eliminates some but not all the defects. If required, start all over again!

A fundamental principle of the scientific method and PDCA is iteration—once a hypothesis is confirmed (or negated), executing the cycle again will extend the knowledge further. Repeating the PDCA cycle can result in getting closer to the goal, usually a perfect operation and output.

In practice, if the change does not work, go through the cycle again with a different plan. If you were successful incorporate in the process what you learned and plan new improvements, beginning the cycle again.

In synthesis, the cycle is:

PLAN ➡	DO ➡	CHECK ➡	ACT ⬌
Develop an organized plan. Identify people responsible for various activities. Describe methods and data to be used to evaluate the effectiveness of the execution of the plan	Test the change. Carry out a small scale study	Check the results. Analyze data. Do they deviate from the plan?	Act on the results. Make improvements and start the cycle again

The PDCA cycle is used when:

- Starting a new project
- Developing or improving a design of a process, or service
- Implementing change in the process
- Working towards continuous improvement

Example

Plan	Design a short course on statistical tools for all analysts
Do	Teach a selected group of analysts
Check	Obtain feedback form the participants
Act	Revise course material, modify the method of teaching etc.
Plan	Redesign the course for presentation to another group of analysts
Do	Deliver the redefined course
Check	Obtain feedback
Act	Act on valuable comments

2.3 Six Sigma Quality Management System

Another well-known version of the Deming model is found in the Six Sigma Quality Management system, which is a problem-solving methodology for improving business and organizational performance.

Six Sigma (*sigma* (σ) is a statistical indication of variation) is a collection of quality improvement techniques that identify the root causes of problems in production or service- delivery processes. It uses quality-analysis techniques and a broad application of statistics to pinpoint the process inputs that cause the undesired outputs. It allows minimizing mistakes and maximizing value and uses the DMAIC (**D**efine, **M**easure, **A**nalyze, **I**mprove, **and C**ontrol) procedure as an iterative cycle. As shown in Fig. 2.3, the steps or components start at the top with Define, then complete the cycle with Measure, Analyze, Improve, and Control.

In this adaptation the:

Define step relates to quality-planning and includes definition of quality objectives and requirements;

Measure this step applies to determine the performance of a procedure, process, or product. Data and tools to understand the cause-effect relationship in the process are used.

Analyze this step involves evaluation of the observed quality, which naturally leads to the Improve step of the process;

Improve during this step, the team get ideas about how to fix the problem in the process. Based on the data collected, the team proposes one or more ways to change the process to eliminate the problem.

Control means maintaining the quality of the improved procedure, process, or product so that it continues to meet the quality objectives and requirements.

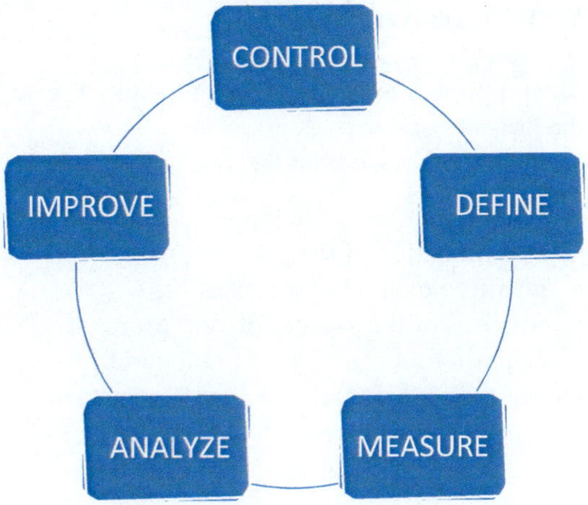

Fig. 2.3 Six Sigma DMAIC Model for Process Improvement

2.3.1 Define, Measure, Analyze, Improve Stages

2.3.1.1 Define Stage

The starting point for the DMAIC problem-solving methodology is the define stage. The key objectives within the define stage of DMAIC are:

- Project Definition: to articulate the project's scope, goal, and objectives; its team members and sponsors, its schedule, and its deliverables.
- Top-Level Process Definition: to define its stakeholders, its inputs and outputs, and its broad functions.
- Team Formation: to assemble a highly capable team and focus their skills on a common understanding of the issues and benefits of the proposed project plans. These objectives should be satisfied before the team progresses to the measure stage.

2.3.1.2 Measure Stage

In the measure stage of DMAIC, a detailed process-level map of the current process is developed. This detailed map clearly defines the activities subject to the improvement efforts. A process consists of repeatable tasks carried out in a specific order.

The objectives of the measure stage include:

- Process definition at a detailed level to understand the decision points and detailed functionality within the process.
- Metric definition to verify a reliable means of process estimation.
- Process baseline estimation to clarify the starting point of the project. A process baseline provides an estimate of the current state of the process and its ability to meet customer requirement
- Measurement system analysis to quantify the errors associated with the metric.

2.3.1.3 Analyze Stage

The purpose is to verify causes affecting the key input and output variables tied to the project goal.

The objectives within the analyze stage of DMAIC include:

- Analyze data collected in measure step
- Determination of the process drivers
- Identify potential root causes
- Analysis of the sources of variation.
- Confirm root causes effect on output
- Collect additional data to verify root causes
- Narrow the research

2.3.1.4 Improve Stage

After understanding the types and sources of variation in the analyze stage, the improve stage can begin to define and implement changes necessary for process improvement.

There are several issues to be completed within the improve stage of DMAIC:

- New process operating conditions are determined.
- Benefits associated with the proposed solution are estimated by the team and approved by the management.
- Failure modes for the new process are investigated and addressed
- Process improvement is implemented and verified.

The outcome of each of these steps may cause a reconsideration of the assumptions from prior steps, forcing further analysis of the proposed improvement. This diligence is often necessary for sustained improvement.

Considering the inherent differences between laboratories based on their resources, skills, and missions, it is expected that Quality Management Systems will vary from one laboratory to another, yet they should all be designed to guarantee that the laboratory fulfills its quality objectives and requirements. It is evident that certain approaches and tools may prove to be useful in many laboratories, but their deployment will depend on the available resources and skills of the laboratory.

What is different about QMS is not the activities or requirements necessary to guarantee the quality of laboratory tests, but how to organize them to ensure quality in the daily management and production of laboratory tests.

Deming's PDCA, or DMAIC Six Sigma models provide the fundamental guidance for structuring Quality Management Systems, but they need to be adapted, expanded, and detailed for efficient and effective management of laboratory testing processes.

In practice, the Quality management identifies the itinerary of activities while a quality system provides a plan for organizing and implementing those activities.

In the next chapter we examine the peculiarities of a quality system in laboratories wishing to get accreditation as required by the ISO17025

2.4 Quality Systems

A quality system refers to the organizational resources, processes and procedures to implement quality management, which is broader than both quality assessment (QA) and quality control (QC). The laboratory quality system, in fact, includes management of equipment, supplies and inventories, management of capital, finances and budgeting, and providing training and continuous support of staff and customer service.

The Quality Assurance (QAssu) program is the backbone of the laboratory quality system and has the purpose of ensuring that the results of a laboratory's analysis are true, accurate, and reliable. In other words, *quality assurance* is *a documented system of protocols to assure the accuracy and reliability of analytical results.*

The relationship between the quality system, QAssu, QA and QC is shown in Fig. 2.4.

2.4.1 Quality System, Assurance, Assessment and Control

Quality assessment (QA) provides broader monitoring of laboratory performance whereas quality control is focused on monitoring the analytical quality of the test results. Quality assurance (QAssu) is the outcome of the entire quality management system.

In practice, the QAssu program incorporates those planned and systematic laboratory activities that guarantee the accuracy and defensibility of testing results. The quality manual, Standard Operating Procedures (SOPs) and documentation are essential components of a QAssu program.

Besides thorough documentation of all procedures and processes, the laboratory also needs to choose the correct methods for testing and establish protocols to detect errors and initiate corrective actions. Validated methods are the technical core of laboratory testing. To determine whether methods are fit for their intended purpose, the selected methods must have established accuracy, precision, calibration and

Fig. 2.4 Relationships in a quality system

limits of detection and quantification. For methods that are taken from a recognized resource, usually a verification process is sufficient to establish the method performance for the lab. However, if the analytical method is developed in-house, a full validation is required.

2.4.2 Principles of Quality Control

Quality control techniques include all practices and procedures that lead to statistical control and to the achievement of the accuracy requirements of the measurement process. The basic element of quality control can be:

- Technical competence of the staff
- Suitable facilities and equipment
- Good laboratory practice
- Good measurement practices
- Standard operation procedures
- Inspection
- Documentation
- Training

QC refers to a measuring process, or to check a result and provide assurance that all activities are performing within predetermined limits. One of the key QC processes in any laboratory is Statistical Process Control (SPC), which utilizes statistical methods to evaluate variability in the laboratory testing and the stability of laboratory procedures. Control charts (see part II) are the most commonly adopted tools to monitor the testing procedures. They are often generated by calculating the long-term mean and range by averaging multiple sets of experimental duplicates over time. By doing such calculations, the laboratory can establish an expected average and variation for future comparison. Control charts thus provide a standard against which the stability of the lab performance can be evaluated.

Other QC procedures that ensure that the laboratory results are of required quality include instrument calibration, use of reference materials, repeated analyses and sample and reagent blank analyses.

2.4.3 Principles of Quality Assessment

Quality assessment techniques consist of ways in which the measurement process may be monitored in order to infer the quality of the data output. They provide assurance that statistical control has been achieved and is maintained as well as estimates of the accuracy of the data.

These techniques can be divided in internal and external. Some of these techniques can be:

Internal	External
Repetitive measurements	Collaborative tests
Internal test samples	Exchange of samples
Control charts	External reference materials
Interchange of operators	Internal reference materials
Interchange of equipment	Audits
Independent measurements	
Definitive method measurement	
Audits	

The components of a QA program include, but are not limited to, the following:

- QA manual
- Staff qualifications and training (initial and in-service)
- Proficiency testing (internal and/or external)
- Sample collection, handling, and storage
- Documented, standardized, and validated procedures
- Reagent and instrument reliability
- Authenticated reference material

Traceability, uncertainty and proficiency testing are the three major items to be addressed in a QA program (see next chapters)

2.4.4 *System Planning*

However, to establish a good quality assurance program, it is essential that senior and/or top management in the laboratory be totally committed to this goal. Management must also communicate this commitment forcefully to the laboratory staff, so that there are no doubt in employees' minds that this is what is expected of them. In other words, all affected departments should be involved in the process. The entire process can be divided into two phases:

1. Investigation phase where an organization programs and develop a quality system "from scratch". It is completed when the newly developed system is in place and operating.
2. Implementation phase. This phase is a long-term, continuing process and involves constant appraisal, review, and planning to update, improve, or correct deficiencies in the system.

2.4.5 *Investigation Phase*

During the investigation phase, Top management collect the necessary information, decides if becoming accredited makes good business sense, and typically follows these following steps:

1. Top Management initiates, funds, and otherwise supports the investigation.
2. Top Management nominates, as project manager, a person that should have experience in laboratory operations, good business sense, an understanding of quality systems and excellent communication skills. The project manager vested with the responsibility and authority necessary to ensure that the system is implemented and maintained.
3. The **project manager** will recruit a **quality manager** as the laboratory representative in charge of quality that will coordinate all the operations for implementing the quality system. The quality manager in turn will involve all staff members, each of whom should be given clearly defined duties and deadlines.
4. The project manager with the support from the quality manager, can decide the **formation of a Steering Team (ST) that** will coordinate the implementation of the project. This team will play an important role in planning, coordinating and providing resources for the ISO 17025 project. The members of the steering team should come from laboratory management, QA, finance, human resources, training, and documentation groups.
5. **The Steering team** defines the scope of the intended accreditation. This could include all calibrations and/or tests performed in a lab, or just part of them. The project team studies the accreditation requirements in detail, having the standard ISO/IEC 17025 as guidelines. The steering team develops a requirements list that should include all documents as required by the standard, for example, policies, a quality plan, and procedures for most of the requirements.
6. **The steering team** can make up **Task teams** for the different tasks required by the quality system.

2.4.6 *Quality Manager Responsibilities*

The quality manager holds the broadest set of responsibilities of anyone in the laboratory. The position is so essential to the quality of a laboratory's analysis that it is specifically required by the ISO 17025 standard.

The quality manager will:

1. Prepare the documentation system;
2. Edit the quality manual (QM) on the basis of the criteria laid down in Standard ISO17025. QM will define the laboratory's quality policy and objectives and outline the quality system framework;
3. Prepare and implement the procedures, i.e. administrative, general and calibration techniques and operation of the equipment;
4. Validate analytical methods and implement the analytical quality control system; and
5. Conduct internal audits to assess whether the quality guarantee system complies with Standard ISO 17025

In addition to focusing on quality aspects, the quality manager will identify what background training is required by different staff members in technical aspects which have to be implemented within the scope of the accreditation process, such as instrument calibration, method validation, calculation of uncertainties, etc.

Another responsibility of the quality manager is to get any useful information from the accreditation agency, control infrastructure and work environment and most important to discuss with the actual chief of the laboratory on the analyses to be accredited. This is very important since this is related with the international methods to be adopted and in turn with the instrumentation availability.

In other words, the quality managers should report to the chief person responsible of the project any problem that can arise in order to take any required the necessary action in time.

2.4.7 Steering Team Responsibilities

The steering team, that can be formed, for instance, by the quality manager and by the chiefs of each laboratory section, will:

- Identify what procedures need to be developed and identify people to work on the procedure; they make up a "Task Team (TT)" (see below) for that procedure;
- Assign start dates and completion dates for each team, and put together, for instance, in a Gantt chart for the project. Teams can be active at different times;
- Evaluate, on a scheduled basis, the progress made, answer questions from the task teams and evaluate resource needs for the implementation; Review, approve or suggest changes to procedures as they are finalized;
- Helps to coordinate the actions of the different Task Teams;
- Identify team members for each procedure;
- Identify training needs for employees and schedule training sessions for ISO 17025:2005;
- Meet on a regular basis to evaluate progress, answer questions for the teams and evaluate resource needs for the implementation;
- Review and approve procedures as they are finalized; and
- Ensure formation of laboratory auditors capable to perform internal audit programs.

2.4.8 Task Team Responsibility

A Task Team should be set-up for each Quality System Procedure that needs to be developed and should be able to train Employees on the basics of ISO 17025. A member of the Steering Team should lead the Task Team. This ensures that the Steering Team addresses questions and issues that arise in the Task Teams.

Furthermore, Task Teams hold team meetings to:

- Review procedure templates and current process;
- Review Gap Analysis results;
- Edit procedure template for your organization;
- Submit the edited procedure to the ISO Steering Team for approval; and
- Take any necessary action to implement the new procedure.

2.4.9 Timeline

Once, goals and responsibilities for the project are established, it is important to:

1. Define starting date of the project. Consider the start of your project as the date of the Gap Analysis (see TIP);
2. Identify one or more people capable to conduct the Gap Analysis. They should have some quality system or audit experience;
3. Define the starting date and schedule of the GAP analysis. A Gap Analysis can typically take from two days to five days to perform. It will depend on the size of the organization, the number of auditors, the state of the current quality system and the experience of the auditors;.
4. Establish dates for introductory training for all employees; and
5. Establish dates for planning meetings

2.4.10 Implementation Phase

Once the decision for ISO/IEC 17025 accreditation is made, and responsibilities defined, the laboratory develops and implements documentation in preparation for the accreditation

assessment.

Typically, implementation follows these steps:

1. The project manager forms task teams for different areas. It is critical that all affected departments at all management levels are represented in the teams;
2. The project management, with the help of QA, searches for an accreditation body and selects the one that best fits the laboratory's needs;
3. The team will train employees that are affected by or have responsibility for the procedure;
4. The teams develop documentation such as procedures under the supervision of the quality manager and send it to the Steering Team for review and approval;
5. Each responsible team will evaluate the process presented in the procedure, determine if any changes are necessary for your organization, and make edits to the procedure and forms;

6. The employees will start following the documented process and maintaining records;
7. The project manager arranges for staff training on the advice of the task teams;
8. Quality assurance performs an internal audit and initiates corrective actions, if necessary;
9. The quality manager prepares the quality manual (SEE TIP);
10. The selected accreditation company performs a pre-assessment;
11. The project manager and quality manger initiates necessary corrective action; and
12. The accreditation company can perform an accreditation audit.

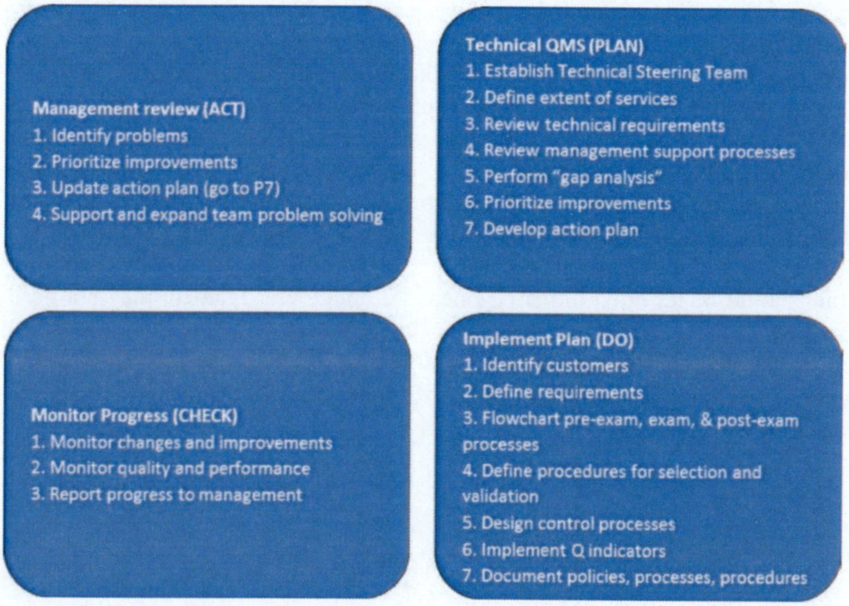

2.4.11 *Consolidating the Program*

As the quality assurance SOPs are written, they should be circulated in draft form to the various laboratory supervisors, and other key technical personnel, for their evaluation and input. When the entire program has been codified, a second meeting should be called for an in-depth discussion of the program.

At this meeting it is important to verify that the written SOPs correspond to the reality of actual laboratory operations. What is necessary is that SOPs are workable, and that laboratory personnel will put them into practice. If there are any complaints about the SOPs, the quality assurance should explain the basic reason for the change in terms of the objectives of the program, and the vulnerability of the laboratory if the change is not made. The point of this whole exercise is to obtain agreement on the part of the technical personnel so they will put into practice the SOPs as they are written. The program should not be imposed and consensus is the key to be successful.

2.4.12 Monitoring and Evaluating the Program

Once consensus has been achieved, set a date for implementing the program. This is conveniently a month or two after the program is in final form, allowing time for educating and training personnel, purchasing any needed material such as notebooks etc., or setting aside areas for archives and sample storage. The quality assurance manager monitors the efforts at complying to the SOPs and when is convinced that most of the program is in place, conducts an audit of the laboratory.

This audit will disclose areas of noncompliance and may require further revision of the SOPs. If an honest effort has been made to comply, and it is found that compliance is difficult or perhaps impossible, it may be necessary to compromise on the requirements of an SOP. (See TIP on auditing). Report all the audit results to the quality manager that will help to interpret results and resolve quality problems.

However, all problems, deficiencies, or irregularities in the quality assurance program should be reported to management. Although management may not need to take direct action to correct problems, as the person(s) with ultimate responsibility for the data quality, they should be made aware of any problems that exist. To ensure that this happens, a formal program for reporting irregularities should be in place.

2.4.13 Management Review

The Quality Manager should prepare for management a "state of the laboratory quality assurance system report" at least annually, or at such other intervals as the needs of the organization may dictate. This document should include reports on activities and topics such as:

- The results of both internal audits and those conducted by an external agency;
- Customer complaints and feedback, if any;
- Initiation and completion of corrective and preventive actions required as a result of audit recommendations;
- Analytical and testing performance quality;
- Follow-up on management orders resulting from previous annual quality system reports and recommendations for improvement; and
- Summary Quality Cost Report for the period since the last annual report.

After reviewing the annual report, management should, prepare the module Corrective Action Requests, in which are described procedures to improve the structure and outputs of the laboratory quality management system. If any recommendations are relative to the acquisition of new equipment, the necessary arrangements should be made to evaluate and determine the validity of such requests.

2.4.14 Communication and Motivation

Finally, effective laboratory management is characterized by an efficient communication system that works both up and down within the organization. Management makes sure that all concerned are aware of quality policies, objectives, plans, and procedures, while seeking out and encouraging timely reports on work progress, improvement, and problems.

Methods for carrying out such communication activities include:

- In-house educational programs related to quality systems;
- Management visits work areas;
- Planned, periodic section meetings; and
- Use of timely and appropriate notices and posters on the laboratory bulletin board.

Furthermore, Top Management must incentivise and motivate personnel to produce results with consistently high quality and continuing quality improvement, which satisfies customer expectations. The management policy of "Total Quality Control," and "Continuous Quality Improvement," must be embraced by all employees. The best way for a laboratory director or supervisor to demonstrate a desire to achieve high-quality results is to show continuous, conscientious, participatory interest in quality activities, especially in the areas of demonstration of quality improvement efforts and meeting customer requirements.

Continuous improvement sounds so simple on the surface, but obstacles do exist. Some of them are reported below:

- **Resistance to change:** Some people may not be happy with the current situation, but they think that any disruption to the process is just as likely to make the problem worse as it is to make it better. It is best to address any of their concerns before making a change. In some cases, any change is considered bad, so a push from the local manager may be necessary to put an improvement in service.

2.5 Tips

2.5.1 Tip: 1 – Human Resources

All personnel involved in any function affecting data quality (sample collection, analysis, testing, data reduction, calibration of instruments, and other quality assurance activities) must have sufficient training in their appointed job to enable them to generate and report accurate, precise, and complete data. The Quality manager has the responsibility for seeing that the required training is available for these personnel and for taking appropriate remedial action when it is not.

Quality control training programs should have the objective of seeking solutions to laboratory quality problems.

This training objective should be concerned with the development, for all laboratory personnel in any aspect or function affecting quality, of those attitudes, that knowledge, and those skills which will enable each person to contribute to the production of high-quality data continuously and effectively.

For instance, certain complex testing (like GMO or pesticides) or analytical techniques (MS/MS) may require specialized training and the formal qualification of technicians or operators. Once individuals are qualified, records must be kept and requalification undertaken periodically, as necessary. The Quality control manager is responsible for the maintenance of such records and for seeing that requalification is accomplished in a timely manner.

A number of training methods are available for laboratory personnel dealing with quality control:

1. Experience training or "On-the-job" (OJT) training is the process of learning to cope with problems using prior experience;
2. Guidance training is OJT with outside help from supervisors or co-workers.

Employees involved in an effective program employing OJT techniques will:

(a) Observe experienced technicians or operators perform the necessary steps in a test or analytical method;
(b) Perform the various operations in the method under the supervision of an experienced technician or operator; and
(c) Perform operations independently but under a high level of quality control checks.

Furthermore, evaluation of the effectiveness of training is accomplished by the conduct of periodic intra-laboratory proficiency testing of laboratory personnel involved in the analytical activity. This evaluation should lead to the determination of the level of knowledge and skill achieved by the technician from the training experiences and the appraisal of the overall training effort, including the discovery of any training areas that show the need for improvement.

2.5.2 *Tip: 2 – Scheduling and Conducting the Gap Analysis*

The project team prepares a gap analysis by comparing ISO requirements with what is already available and implemented. A gap exists where existing policies, processes or procedures do not fully meet the stated requirements. This analysis should include all processes and procedures for management controls and technical controls, such as for sampling, method validation, equipment calibration, qualification and maintenance, employee qualifications, and others.

Using the outcome of the gap analysis, the project team develops a task list. The list is completed with additional tasks such as selecting and dealing with an accreditation body.

The project team, maybe with the help of an external consultant, makes an estimation of the overall ISO/IEC implementation costs, which should include costs for initial set-up and for maintaining the quality system. The costs are compared with the direct and tangible estimated additional returns that come from getting accreditation status. Tangible returns are, for example, savings through more efficient operation.

The team makes a rough estimation of the return on investment for both the short and long- term views, and makes a recommendation to management. Management decides to accept or reject the proposal.

A. Schedule the Gap Analysis

(a) Identify person to conduct the gap analysis.
(b) Fix dates and communicate to all employees the audit schedule. In the note specify what is being done, by whom, and why.
(c) Determine if the audit will be by process or by area of the facility. Usually, the audit by area of the facility is preferred.
(d) Divide the facility into manageable areas. Schedule a time of audit for each section
(e) Assign the team to cover the various areas of the facility.
(f) Arrange the Gap Analysis checklists so each auditor will have the sections of the standard that are applicable in the areas they will cover.

B. Conducting the Audit

(a) Follow the programmed schedule. Visit each area of the facility to evaluate the current quality system. Focus on what is in place, and what is not in place.
(b) Take notes on what is in place, and what will need to be developed and changed. Take complete notes, reference documents and examples.

C. Reporting

(a) Summarize the audit findings in the form of a task list.
(b) Identify the status of the current system for each requirement (or set of requirements) of the ISO 17025.
(c) The Steering Team will use this information to assign responsibilities and timelines to task teams.

2.5.3 *Tip: 3 – Quality Manual*

The Quality Manual is the master document for laboratory accreditation and its purpose is to describe the quality system being developed and implemented as well as methods adopted to ensure its fulfillment to ISO17025. It serves as the primary resource for laboratory information and is the culmination of a planning effort to design into a program or specific project provisions and policies necessary to assure accurate, precise, and complete quality data.

The Quality Manager should draw up or co-ordinate the Quality Manual and should also determine, for each requirement, what documentation is needed and create working teams representing different departments.

Top management should write the quality policy statement, which should outline the laboratory's commitment to quality. The Quality manger is responsible for ensuring adherence to the laboratory quality manual and Standard Operation procedures (SOPs).

The standard operating procedures (SOPs) or work instructions give step-by-step instructions on performing tasks. Examples of SOPs are procedures for checking and calibration of equipment. All laboratory SOPs should use the same format, to make writing and reading easier. A good practice is to have an SOP for how to author, review, approve, distribute, and update SOPs. Preferably, senior members of anticipated user groups should write SOPs. This helps ensure that SOPs have the right level of information and are used and followed.

2.5.4 *Tip: 4 – Example of Task Assignments*

Task assignment	Team members	Team leader	Start date	End date
Quality manual				
Document control				
Training				
Infrastructure control				
Analytical processes				
Purchasing				
Identification and traceability				
Preservation of product				
Control of measuring and monitoring devices				
Internal audits				
Control of nonconforming products				
Corrective action and preventive actions				
And so on				

2.5.5 *Tip: 5 – Example of Project Gantt Chart*

(Identify when each team will start and stop the duty and shade the time that each team will run)

Task group	Month 1	Month 2	Month 3	Month 4	Month 5	Month 6	Month 7	Month 8	Month 9
Management responsibility									
Training									
Infrastructure									
Purchasing									
Traceability									
Sample preservation									
Instrumentation control									
Internal audits									
Validation									
Processes analysis									
Nonconforming product control									
Corrective/Preventive actions									
And so on									

Chapter 3
Preparing for Analysis: The Analytical Method

Methods are the core of the analytical process since they report a set of instructions to be followed for obtaining reliable results. The selection and development of methods of analysis has traditionally been a subject of importance to analytical laboratories working in the food sector. For instance, organizations, national or international, working in the foodstuffs area, develop methods of analysis and incorporate them into legislation.

Nevertheless, it is essential that laboratories ensure the quality of all their analytical methods in use by careful verification and validation.

The requirement for food analysis laboratories to use a 'fully validated' method of analysis is now universally accepted or required within the food sector.

Consequently, a 'validation plan' should be written that indicates the method criteria needed and addresses questions such as:

- when is the method going to be used? (official food control and in-house process control methods may have to fulfil different criteria on, *i.e.* precision and accuracy)
- what type of answer is required: qualitative or quantitative?
- in what state is the analyte, *i.e.* is it bound or free?

It is then important for a laboratory to choose a fit for purpose method of analysis.

3.1 Sources of Methods

Selection of a suitable analytical method can be made once the reason for carrying out the analysis is well understood. Analytical methods may be (a) qualitative or (b) quantitative or semi-quantitative. The former usually pose few problems if only an indication is required as to whether a particular analyte is present or not. If a

S. Mannino, *Guidelines for Laboratory Quality Managers*, Integrating Food Science and Engineering Knowledge Into the Food Chain 14,
https://doi.org/10.1007/978-3-031-11724-4_3

negative result is required (i.e. confirmation of absence from the product), then one has only to worry about the limit of detection of the test used. Equally, rapid tests for positive confirmation are often made on unknown substances. Quantitative methods are used in many situations and a variety of different methods can be employed. Prerequisite for a successful selection is that the purpose of the analysis. Laboratories carrying out analytical work for official food control do not always have the freedom to select the method. In particular, microbiological methods are often laid down in the legislation, *i.e.* both EU and national legislation frequently specify which reference methods are to be used in cases of dispute.

In some cases, laboratories use internally (in-house) developed methods or significantly modify standard methods. Such methods must be validated at the relevant concentration ranges, before being taken into routine use.

If the method of choice is an established standard reference method from, for example, the **AOAC** International, the laboratory usually only needs to verify that it can achieve the performance characteristics given in the method, especially trueness and precision, and demonstrate that the method is suitable for the intended use.

However, the introduction of new analytical methods requires that the laboratory is sufficiently competent, experienced and proficient.

There are many types of routine analysis carried out by laboratories, i.e. those required for screening purposes, where one is testing a large number of individual unrelated samples, and those for surveillance activities, e.g. monitoring foodstuffs for the level of toxic metals. **Screening methods** must be extremely rapid, permitting a high throughput of samples at low cost. These methods can be qualitative or semi-quantitative, and may be validated only to the extent of the limit of detection, by the operational laboratory.

Surveillance methods are very similar to screening methods but are usually somewhat less rapid, with a lower throughput of samples although they do yield quantitative results.

Other methods can be:

Regulatory and reference methods are typically identified by an official body for use in enforcement of a specific regulation. Although these methods will have been fully validated, the laboratory using the method has to verify that it can achieve the published performance requirements. **Reference methods** will also have been fully validated and tested by an approved collaborative study. These types of method are frequently used to characterize reference materials.

Primary methods have the highest metrological qualities, whose operation can be completely described and understood, and for which a complete uncertainty statement can be written down in terms of SI units.

However, factors that have to be considered when choosing between types of method should include:

- Type of sample, matrix and measurand;
- Time and cost;
- Availability of equipment;

- Frequency of false negatives and false positives;
- Limit of detection and quantitation;
- Working range;
- Selectivity;
- Quantitative or semi-quantitative;
- Acceptable uncertainty.

3.2 Evaluation of Published Methods

Laboratories seeking accreditation for chemical, microbiological and sensory food analyses must be able to demonstrate that they can competently use the methods included in the scope of the accreditation. If a method is to be used for the official control of foods there are extensive requirements on internal verification, ***i.e.*** that the laboratory is able to demonstrate that it can use the method in a way which enables the analytical task to be solved.

The methods selected should be chosen on the basis of their practicability and preference should be given to official methods elaborated by international organizations and which have applicability for routine use.

Suitable methods may be found:

- published in the open scientific literature, e.g. *The Analyst*, *Journal of AOAC International*, *Journal of Chromatography*, *Electroanalysis*, *Journal of AOACS and so on*
- from national or international organizations, e.g. (UK) BSI, BP; (International) ISO; (Europe) CEN, EP; (USA) ASTM, EPA, USP, etc.

However, laboratories for food quality control use generally well established methods of analysis developed by international organization. These are:

3.3 AOAC International (AOACI)

AOAC INTERNATIONAL is the forum for finding appropriate science-based solutions through the development of microbiological and chemical standards. AOAC's primary activity is the development of globally accepted standards. AOAC standards are used globally to promote trade and to facilitate public health and safety.

AOAC develops analytical methods for a broad spectrum of safety interests including;

- foods and beverages
- dietary supplements and infant formula
- feeds
- fertilizers

- soil and water
- veterinary drugs
- pharmaceuticals
- and more

It also provides Laboratory Proficiency Testing Program, accredited by A2LA.

AOACI requires that all of its methods, which are used on a worldwide basis, be collaboratively tested before being accepted for publication in the organization's 'Official Methods of Analysis' book; it now adopts only methods which have been validated in a collaborative trial.

3.4 The Codex Alimentarius Commission

The C O D E X A L I M E N T A R I U S international food standards, guidelines and codes of practice contribute to the safety, quality and fairness of this international food trade. Consumers can trust the safety and quality of the food products they buy and importers can trust that the food they ordered will be in accordance with their specifications.

This was the first international organization working at the government level in the food sector, which laid down principles for the establishment of its methods of analysis.

These methods are primarily intended as international methods for the verification of provisions in Codex standards. They should be used for reference, in calibration of methods or introduced for routine examination and control purposes.

Useful information about methods of analysis and sampling can be found at the following website www.codexalimentarius.org/input/.../388/CXS_234e_2014.pdf

3.5 The European Union

The European Union (EU) is attempting to harmonize sampling and analytical procedures to meet the current demands of the national and international enforcement agencies.

The criteria are similar to those recommended by any international organization and criteria are can summarize in:

- Specificity
- Accuracy
- Precision; repeatability intra-laboratory (within laboratory), reproducibility
- inter-laboratory (within laboratory and between laboratories)
- Limit of detection, Sensitivity
- Practicability and applicability under normal laboratory conditions
- other criteria which may be selected as required.

Useful information can also be found in the following document of EU available for free on Internet:

Regulation (EC) No 882/2004 of the European Parliament and of the Council of 29 April 2004 on official controls performed to ensure the verification of compliance with feed and food law, animal health and animal welfare rules is the legal framework for sampling methods and methods of analysis of feed for control purposes.

Commission Regulation (EC) No 152/2009 of 27 January 2009 laying down the methods of sampling and analysis for the official control of feed establishes the sampling method and the methods of analysis of feed for control purposes.

In 2013, the sampling procedure has been completely updated by Commission Regulation (EU) No 691/2013 of 19 July 2013 amending Regulation (EC) No 152/2009 as regards methods of sampling and analysis.

A guidance document for the implementation of Commission Regulation (EU) No 691/2013 has been elaborated and has been endorsed by the Standing Committee on the Food Chain and Animal Health – section Animal Nutrition.

A List of Official EU Methods of Analysis can be found at the following website: http://www.agriculture.gov.ie/agri-foodindustry/feedingstuffs/samplinganalysis/listofofficialeumethodsofanalysis/

3.6 The European Committee for Standardization (CEN)

The European Committee for Standardization (CEN) is an international organization, based in Europe, which is, now developing most standardized methods of analysis in the food additive and contaminant areas and in some commodity specific areas. CEN's activities in relation to food safety are in line with the European Union's objective to achieve the highest possible level of health protection for the consumers of Europe's food.

EU food safety legislation establishes a cascade of methods that shall be used for official control purposes. Preference is given methods that comply with internationally recognized rules or protocols, like those described in CEN publications. Therefore a majority of European Standards and other deliverables developed by CEN in the area of Food and Feed are supported by Mandates from the European Commission requesting development of validated methods of analysis of food and feed.

However, CEN methods are not prescribed by legislation, but the European Commission does place considerable importance on the work that CEN carries out in the development of specific methods in the food sector; CEN has been given direct mandates by the Commission to publish particular methods, ***e.g.*** those for the detection of food irradiation.

CEN, like the other organizations described above, has adopted a set of guidelines to which its methods technical committees should conform when developing a method of analysis.

In addition, CEN publishes its methods either as finalized standards or as pre-standards, the latter being required to be reviewed after two years and then either withdrawn or converted to a full standard. Thus, CEN follows the same procedure as the AOACI in this regard.

Besides European Standards, CEN produces other reference documents, as: Technical Specifications, Technical Reports and Workshop Agreements.

3.7 ISO

ISO (International Organization for Standardization) is an independent, non-governmental membership organization and the world's largest developer of voluntary International Standards.

It counts the adhesion of 162 member countries who are the national standards bodies around the world, with a Central Secretariat that is based in Geneva, Switzerland.

ISO has published more than 19,500 International Standards covering almost every industry, from technology, to food safety, to agriculture and healthcare. ISO International Standards influence everyone, everywhere.

ISO International Standards ensure that products and services are safe, reliable and of good quality. For business, they are strategic tools that reduce costs by minimizing waste and errors and increasing productivity. They help companies to access new markets, level the playing field for developing countries and facilitate free and fair global trade.

Information and methods of analysis can be found at the following web site in English http://www.iso.org/iso/home/about.htm

Nevertheless, before taking new or unfamiliar methods into routine use, authorized staff (as quality manger) must decide if the requirements of the customer and of the analytical problem are being met. The judgement should be based both on published validation data and on experimental data resulting from the laboratory's own verification work.

Standard methods from organizations which develop methods are often clearly worded and can therefore be used without modifications.

In some cases, however, methods contain alternative procedures for extraction or other analytical steps. In such cases, quality manager or lab chief, must specify which alternative is to be used. If it is necessary to introduce modifications, these must be clearly documented, dated and signed by authorized staff. The laboratory shall maintain records of all authorized instructions, *etc.*, and the instructions must be included in the document control system of the laboratory.

When no validation data are available, then all of the relevant parameters will have to be studied. The degree of rigor with which the study is carried out will depend on issues such as 'criticality' of the measurement and the availability of validation data on similar methods.

It is important that any method chosen is scientifically sound under the conditions it will be applied. It is also necessary to demonstrate that the equipment, which will be used, is suitable and its use will not influence the results adversely.

This includes all types of equipment, from volumetric glassware to instruments that should have sufficient sensitivity over the entire range of measurement.

Chapter 4
Statistics for the Quality Control Laboratory

To verify any process and express the results in a form required by customer, a set of statistical procedures must be present along the chain of the analytical process.

This chapter contains a brief description and worked examples of basic statistic procedures that can be helpful in quality control programs.

Although this chapter is a comprehensive, actionable description of methods and tools frequently used in laboratory, we recommend refereeing to authoritative books such as:

- James *Miller*, Jane C *Miller: Statistics* and Chemometrics for Analytical Chemistry (6th Edition)
- *Anderson*, Robert L.: Practical *Statistics for Analytical* Chemists, Van Nostrand Reinhold, 1987 New York
- Stephen L R Ellison, Vicki J Barwick, Trevor J Duguid Farran: Practical Statistics for the Analytical Scientist: A Bench Guide: Edition 2

The field of statistics is typically divided in two areas of study:

1. **Descriptive statistics** that represent a characteristic of a large group of observations (a population or a sample of population). Example: mean and standard deviation are descriptive statistics about a set of data).
2. **Inferential statistics** draw conclusions about a population based upon analysis of sample data. A small set of numbers (a sample) is used to make inferences about a much larger set of numbers (the population). This type of statistics will be used in hypothesis testing.

S. Mannino, *Guidelines for Laboratory Quality Managers*, Integrating Food Science and Engineering Knowledge Into the Food Chain 14,
https://doi.org/10.1007/978-3-031-11724-4_4

4.1 Data Presentation

In most cases, the data used in the laboratory is a sample (a subset) taken from a population and parameters are terms used to describe the key characteristic of the population.

However, before any statistical calculations can be performed on data, it is necessary to arrange them in some way in order to have a clear picture of their features. This can be done in a qualitative way by grouping the measurements, by forming frequency tables and charts, or by using descriptive statistics.

An example of raw data would be:

3.65	3.15	3.03	3.05	3.35	3.65	3.15	3.35	3.15	3.67

Data can be arranged in order of magnitude, in which case it is referred to as ranked data or sorted data. In this form, simple inspection will disclose such features as steady progression, a hump, a gap, or a biased frequency distribution. The same data as above, ranked, is as follows:

3.03	3.05	3.15	3.15	3.15	3.35	3.35	3.65	3.65	3.67

For instance, in this way is easy to notice the higher and smaller value and to calculate the range (see below). The same data may have grouped into intervals or presented graphically as a bar chart, box plot or histogram. All these functionalities are available on the popular spreadsheet Excel.

4.2 Measure of the Central Tendency (Mean, Median, Mode)

Central tendency shows how tightly data cluster around the central point. The three most common measures of the central tendency are mean (or average) median, and mode, the most commonly used being the arithmetic mean.

The mean is the sum of the individual values in a set of data divided by the number of values. In our example:

$$\overline{x} = \frac{\sum_{i=1}^{n} x}{n} = 0.32$$

4.2.1 Median

Median is the midpoint of a ranked order set of data and is often a more meaningful location parameter than the mean. In fact, it is not much affected as much by outliers. The median is the center value if there is an odd number of data point or the average of two middle values in case of even number of data points. In our example, the median is 3.25 (average of 3.15 and 3.35).

4.2.2 *Mode*

The mode of a set of data is the most frequently observed value. In our example 3.15, value that occurs three times.

4.3 Measures of Spread (Range, Variance, Standard Deviation)

In addition to the location parameters is important to know the spread or the degree of scatter of the individual values. Spread tell us how the data are distributed around the center point. Higher the spread, higher the variation. Common measure of spread include range, variance, an standard deviation.

4.3.1 *Range*

Range is the difference between the largest and the smallest values in a data set.

In our example: range = R = (3.67–3.03) = 0.54.

4.3.2 *Variance*

The variance shows how far off the data values are from the mean overall.

Mathematically, the variance is the sum of the squares of the differences between the individual values and the arithmetic mean of the data set, divided by the number of data values minus 1 (n − 1).). The divisor (n − 1) is used, rather than n, because the sample mean is known, and all but one of the values, is possible to calculate what that last value must be. Statisticians say there are n − 1 degrees of freedom.

For the sample

$$s^2 = \frac{sum\ of\ squares}{n-1} = \frac{\sum_{i=1}^{n}\left(x_i - \overline{x}\right)^2}{n-1}$$

while for the population

$$\sigma^2 = \frac{sum\ of\ squares}{n-1} = \frac{\sum_{i=1}^{N}\left(X_i - \overset{?}{\overline{x}}\right)^2}{n-1}$$

Generally, in statistics, parameters are the terms used to describe the key characteristics of a population. The population parameters are denoted with a small Greek letter such as σ (standard deviation) or μ (average). From the analytical point of view, a population is a collection of all possible analytical results from a given lot of material using a given chemical method. For a sample drawn from a population, Latin letters (Z, s, x) are used to describe the sample statistics. The information that we collect from the sample units is used to estimate population parameters.

In our example the variance is

$$Variance = 0.0651$$

The variances are additives while the standard deviation not.

4.3.3 Standard Deviation

The standard deviation is the square root of the sample variance and can taut as the average distance of each point to the mean. For a sample statistic:

$$s^2 = \sqrt{\frac{sum\ of\ squares}{n-1}} = \sqrt{\frac{\sum_{i=1}^{n}\left(X_i - \overset{?}{\overline{x}}\right)^2}{n-1}}$$

In our example

$$Standard\ deviation\ s = 0.2551$$

Another useful parameter is the relative standard deviation (RSD), also called coefficient of variation that is the ratio of the standard deviation to the mean. If multiplied by 100 gives the relative standard deviation percentage.

In our example RSD is

$$RSD = \frac{s}{mean} = \frac{0.2551}{3.32} = 0.0768$$

And in percentage

$$RSD\% = 0.0768?100 = 7.68\%$$

4.4 Normal Distribution

The most widely used continuous frequency distribution is the Gaussian or nor normal curve such as the one depicted in Fig. 4.1.

In many analytical situations, data follow a normal distribution (bell-shaped curve). One of the key properties of the normal curve is the relationship between shape of the curve and the standard deviation (σ for population and s for sample).

As it can be seen, 99.73% of the area under the curve is contained between ±3 standard deviations.

As the standard deviation increases, the curve becomes more spread out and the frequency density around the mean decreases. Conversely, when the standard deviation decreases, the curve becomes narrower and the frequency density around the mean increases. (See Fig. 4.2).

The area under the normal curve between two ordinates expresses the probability that a random, unbiased measurement from a normal population will fall in the interval bounded by the two points.

The standard normal distribution can be described by a normal random variable Z, that has mean of zero and a standard deviation of 1, where Z is the number of standard deviations a specific observation lies from the mean:

$$Z = \frac{x - \mu}{\sigma}$$

A more rigorous approach to this subject can be found in specialized book. However, some point should be highlighted:

- Many statistical tests or inferences apply only if data are normally distributed
- Normality is required for many statistical test so it is useful to convert no-normal data into something that does have a normal distribution (i.e. in log)

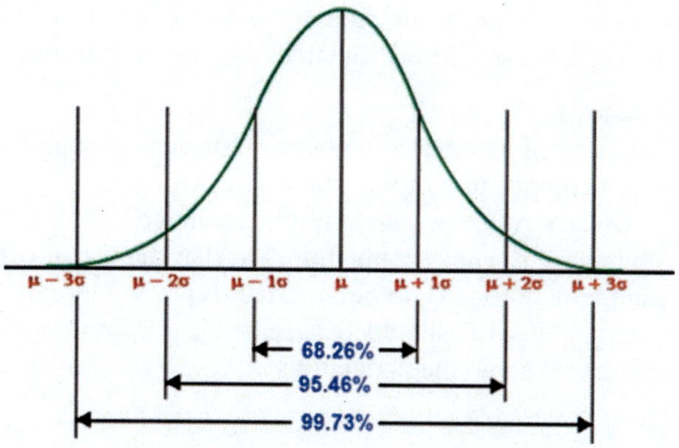

Fig. 4.1 Normal or Gaussian curve

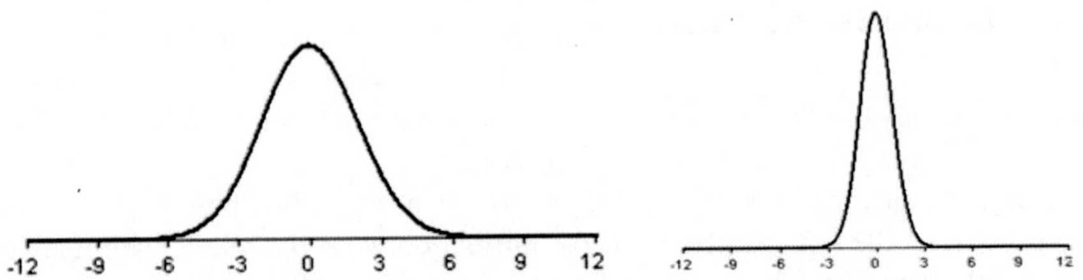

Fig. 4.2 When the spread increase (larger standard deviation) the Gaussian curve gets larger (A) while when decrease (smaller standard deviation) the curve become sharper

- The distribution of averages approach to normality for large samples. This property is called Central limit theorem: The distribution of means calculated from samples approach to normal distribution.
- Calculating averages on subset of data is a common practice when in presence of a non-normal distribution.
- The mean of a number of observations is a more reliable estimate of the population mean than is a single observation.
- The standard deviation of a series of mean values, is much smaller than the standard deviation of a series of individual measurements

4.5 Using Samples to Estimate Population Values

In research, we often want to gather data about a population that is of interest to us. In most cases, however, it is not possible, or not practical, to study all the members of the population. Instead, a sample can be selected from that population and used sample's numerical data (the sample statistics) to estimate the population values (the parameters).

In using statistics to estimate parameters, we can expect some sampling error. Sample values (e.g., the mean) are likely to be higher or lower than the fixed population values they are designed to estimate. Nevertheless, information from a single sample to estimate the parameters of the population from which that sample was selected, is used. This estimate is expressed in terms of probability, not certainty.

Example
Selection of a discrete number of tuna cans (a sample) in a store to study the nutritional composition of the whole stock of cans.

Clearly, when we use sample values to estimate population values, instead of studying the whole population, we risk making a certain level of error. In fact, a sample is not likely to be a perfect representation of the population and a certain "margin of error" should be accepted if is important to use this information to make inferences about the population.

If we select multiple samples of the same size from the same population, compute the sample means, and plot the sample means, we would see that they are normally distributed in a bell-shaped curve.

As the number of samples increases, the shape of the distribution gets closer to a smooth-looking bell-shaped curve.

In the example of the cans, if the analyst want to assess the average (i.e., mean) of the 5.000 cans present in the store, he can select several samples of cans, each with 10 cans, and test their fat content. Almost surely, he will not get the exact same fat content but he can predict the population mean and the standard deviation. The mean of the sample is used to estimate the population mean and the sample standard deviation to estimate the standard deviation of the population.

4.6 Standard Error of the Mean

The standard deviation of the sample means is called the **standard error of the mean**, and is used to estimate the sampling error. This value expresses the amount of variability among the various sample means. It can be calculated from the standard deviation and the number of observations n.

When based on a very large, known population, the standard error is:

$$\sigma_x = \frac{\sigma}{\sqrt{n}}$$

When estimated from a sample drawn from very large population, the standard error is:

$$\sigma_x = \frac{s}{\sqrt{n}}$$

The dispersion of sample means decreases as sample size is increased.

In the formula used to compute the standard error of the mean, the square root of the sample size is used as the denominator. Therefore, higher sample sizes would result in lower standard errors of the mean.

The standard error of the mean tells us that if we were to continue to draw additional samples of the same size from the same population, we could expect that of the sample means:

- 68% would be within ±1SD of our obtained sample mean;
- 95% would be within ±2SD of the obtained sample mean;
- 99% would be within ±3SD of the obtained sample mean.

4.7 Shapiro-Wilks for Testing Normality

The Shapiro-Wilks Test is a hypothesis test that is widely used to determine whether a data sample is normally distributed. As with any hypothesis test, the output states whether to reject or not reject the Null Hypothesis and the specified alpha level. In this case, the Null Hypothesis states that the data sample came from a normally distributed population. This Null Hypothesis is rejected only if Test Statistic W is less than the critical value of W for a given alpha level and sample size.

The test gives you a W value; small values indicate your sample is *not* normally distributed (you can reject the null hypothesis that your population is normally distributed if your values are under a certain threshold). The formula for the W value is:

$$W = \frac{\left(\sum_{i=1}^{n} a_i x_i\right)^2}{\sum_{i=1}^{n} x_i - x_{av)}\,)^2}$$

where:

x_i are the ordered random sample values and a are constants found in the Shapiro – Wilks table.

A test statistic W is calculated. If this test statistic is less than a critical value of W for a given level of significance (alpha) and sample size, the Null Hypothesis which states that the sample comes from a normally distributed population is rejected. Keep in mind that passing a hypothesis test for normality only allows one to state that no significant departure from normality was found.

The Shapiro-Wilk Test is a robust normality test and is widely-used because of its slightly superior performance against other normality tests, especially with small sample sizes. Superior performance means that it correctly rejects the Null Hypothesis that the data are not normally distributed a slightly higher percentage of times than most other normality tests, particularly at small sample sizes.

Steps of the Shapiro-Wilk normality test to be followed:

- Arrange Data in an Ascending order
- Arrange the sorted data into pairs of successive highest and lowest data values
- Calculate the difference between the high and low value of each pair
- Multiply each difference by an "a Value" from a lookup table
- Calculate test statistic W
- Compare test statistic W with a W Critical Value obtained from a lookup table
- Reject the Null Hypothesis stating normality only if W is smaller than W Critical

In more detail: Stp1 - Calculate SS by the formula

$$SS = \sum_{i=1}^{n} \left(x_i - x_{av}\right)^2$$

where SS is the sum of the squared deviations of x from the mean x.

Step 2 – Set the alfa level, 0.05 or 0.10 (w values are provided only for certain alfa level)
Step 3 – Obtain a weights from the Shapiro-Wilk table (based on the values of n). These values are for samples varying in size from n = 2 up to n = 50.The number of a values depends on the sample size n. If n is an even number, the number of a Values equals n/2. If n is an odd number, the number of a Values equals (n – 1)/2.
Step 4 – Pair Up Successive Highest-Lowest Data Values

Create the Upper Value of Each Data Pair
Create the Lower Value of Each Data Pair

Step 4 – Calculate the Difference Within Each Pair
Step 5 – Calculate a x Difference
Step 6 – Calculate Test Statistic W
Step 7 – Lookup W Critical Values

Each unique combination of sample size and alpha level has its own critical W value. The following is a table of the respective critical W value for each combination of n and alpha up to a sample size of 50 and 3 common levels of alpha.

Step 8 – Determine Whether or Not to Reject Null Hypothesis by Comparing W to the critical W of the table

The Null Hypothesis is rejected only if Test Statistic W is smaller the critical W value for the given sample size and alpha level. The Null Hypothesis states that the sample came from a normally distributed population. The Null Hypothesis is never accepted but only rejected or not rejected. Not rejecting the Null Hypothesis does not automatically imply that the population from which the sample was taken is normally distributed. This merely indicates that there is not even evidence to state that the population is likely not normally distributed for the given alpha level. As with all hypothesis tests, alpha = 1 – level of certainty. If a 95-percent level of certainty is required to reject the Null Hypothesis, the alpha level will be 0.05.

The test has limitations, most importantly that the test has a bias by sample size. The larger the sample, the more likely you'll get a statistically significant result.

It's rare to calculate the Shapiro-Wilk by hand. Many software packages including Excel can make these calculations. However, a manually worked example will clarify the Shapiro-Wilk test.

4.7.1 Sulphur Dioxide SO_2 in White Wine

Twelve white wine samples are analysed for their content of Sulphur dioxide. The following table can be prepared.

Test for normality – Shapiro-Wilks n = 12					
	Raw data	Sorted low to high	Differences	From the table	a × differences
1	51	21	$x_{12} - x_1 = 72–21 = 51$	$a_1 = 0.5475$	27.9225
2	47	31	$x_{11} - x_2 = 60–31 = 29$	$a_2 = 0.3325$	9.6425
3	49	41	$x_{10} - x_3 = 58–41 = 17$	$a_3 = 0.2347$	3.9899
4	72	44	$x_9 - x_4 = 56–44 = 12$	$a_4 = 0.1586$	1.9032
5	56	47	$x_8 - x_5 = 51–47 = 4$	$a_5 = 0.0922$	0.3688
6	41	48	$x_7 - x_6 = 49–48 = 1$	$a_6 = 0.0303$	0.0303
7	44	49			S = 43.886
8	31	51			$SS = S^2 = 1925.98$
9	48	56			
10	21	58			
11	60	60			
12	58	72			
	Average	48.16			

Calculate the Sum of squares by squaring the sum of the differences SS = 1925.98

Calculate the sum of the square of the differences of each data and the average (48.16) and obtain the value of B = 1979.72

$$SS = \sum_{i=1}^{n} \left(x_i - \overline{x} \right)^2$$

Calculate W = 1925.98/1979.72 = 0.9728

This value si compared with the value of the Shapiro-Wilks table for n = 12 and p=0.05. The value is 0.859.

Considering that W is larger than that calculated, we accept the Null Hypothesis. Shapiro-Wilk table can be found in the Appendix at the end of the chapter.

4.8 Confidence Intervals

As was explained, using statistics derived from a single sample to estimate the population parameters is likely to result in a certain level of error. A **confidence interval** is a way to estimate the population value that is of interest to us. The confidence interval (CI) lets us predict, with a certain degree of confidence, where the population parameter is. Common confidence interval are 90%, 95%, and 99%. The limits defining the confidence interval are called confidence limits.

The sample mean is used as the center of the confidence interval. While the standard error of the mean is used in constructing confidence intervals. Computer programs usually provide the calculation of the confidence limit by using the following formulas:

$$\mu = \overline{x} \pm \frac{ts}{\sqrt{n}}$$

where symbols have the usual meaning except t (a tabulated value reported in Appendix at the end of the chapter) that will be discussed later.

Example

Given $\dot{x} = 108$, $s = 15$ and $n = 26$, estimate a 95% confidence interval for the population mean. Since the population variance is unknown, t-distribution is used. The resulting interval, using a t-value of 2.060 from the table, is approximately 102–114 ($\mu = 108 \pm 6$). Consequently, any hypothesized μ between 102 and 114 is tenable based on this sample. Any hypothesized μ below 102 or above 114 would be rejected at the 0.05 significance.

4.9 Steps in the Process of Hypothesis Testing

In analytical practice, the analyst has to make decisions about the entire population based on data obtained on a sample of the population. To this end, we formulate hypotheses (statistical hypotheses) whose authenticity must be verified.

Very often we formulate a statistical hypothesis for the sole purpose of rejecting it. For example if we want to know if a method is better than another, we formulate the hypothesis that there is no difference between the two methods. This kind of assumption is called the null hypothesis and is generally indicated by H_0. Any other case other than the null hypothesis is called alternative hypothesis and is denoted by H_1.

Statistical tests of hypothesis testing are rules that allow us to accept the hypothesis or reject it. These type of tests are called significance tests.

Example

Suppose the analyst wants to make sure that the two means (m_a and m_b) obtained by analyzing the same sample several times by two different methods, are the same. The analyst formulates the null hypothesis that the two means are equal and for this, and will use the notation:

$$H_0 : \mu_a = \mu_b$$

The alternative hypothesis can be formulated in three different ways:

1. H1: $\mu_a \neq \mu_b$
2. H1: $\mu_a < \mu_b$
3. H1: $\mu_a > \mu_b$

If the analyst expresses the first 'alternative hypothesis means that he is interested in knowing if the two means are different, in the other two cases he is interested in knowing which of the two averages is the largest or the smallest.

How to formulate the hypothesis determines the type of test to use. In fact, in the first case the analyst use a so-called two-tailed tests because we he is only interested to check whether there is a real difference between the two averages. In the other two cases, it will use the one-tailed test because he is interested in knowing if the average m_b is greater or smaller than m_a.

Because Ho can take any result, every time the analyst reject the null hypothesis, takes the risk of making an error of first kind and whenever he accept the null hypothesis a type II error. In other words, the test allows the analyst to make a decision but with a certain probability of being wrong.

There are four possible outcomes of a test of the hypothesis according to the following schedule:

		True status of H_0	
		H_0 True	H_0 False
Decision	Accept H_0	Correct (1-α)	Type II error (β)
	Reject H_0	Type I error (α)	Correct (1-β)

Type 1 or α error – the rejection of (H_0) when it is actually true.
Type II or β error – the acceptance of (H_0) when it is actually false

When trying a certain assumption, the maximum probability with which we accept the risk of failure of the first type is called the level of significance of the test. In practice, significance levels of 0.05 and 0.01 are used. For example, if for a test of hypothesis significance level 0.05 (i.e. 5%) was chosen, it means that there are 5 chances out of 100 (or 1 to 20) to reject the hypothesis whereas it should be accepted. In other words, we are 95% confident to have made the right decision.

In order to avoid confusion, the analyst should keep in mind that all the following expression have the **same** meaning:

- The result is significant at 0.05 level
- 95% confidence level
- Type 1 error is 0.05
- Significance level α is 0.05
- $\alpha = 0.05$
- Type α error is 0.05

- There is 95% certainty that the result is not due to chance
- There is a probability of 1:20 to obtain this results
- The rejection area is 0.05
- The p value is 0.05
- $p = 0.05$

In practice, once the research question has been formulated, the significance test requires:

- Definition of the null (H_0) and alternative (H_1) hypothesis
- Decision on the significance level of test
- Determining the statistical test(s) to be used to analyze the data.
- Calculating the appropriate test statistics and verification if the results are statistically significant,
- Comparison of the results with the critical tabulated value at the chosen confidence level.
- If, from the comparison, the found value is inferior to that tabulated, the null hypothesis is accepted, otherwise is rejected.

4.10 Example of Statistical Tests Routinely Applied in the Analytical Laboratory

Below some useful statistical tests applied to the analytical data that can help the analyst in making decisions, are reported.

They concern:

- Outliers
- Precision of data
- Averages
- Calibration

4.11 F-Test

4.11.1 Comparison of Two Standard Deviations: The F-Test

One statistical test for significance was developed by Mr. Fisher, and is therefore called the F -test. The test is devoted to evaluating data to determine whether one population is more variable than another. It can be used to determine whether or not two methods differ in their precision, or whether two analyst or instrument or laboratory produces more precise results than another. In the comparison of 2 standard deviations, the statistic F is calculated as follows

$$F = \frac{s_a^2}{s_b^2}$$

where s_a^2 = is the higher of the two variance with na-1 degree of freedom for population A and s_b^2 = the second estimated variance with nb-1 degree of freedom for population B.

The observed value of F is compared with Fa (dfl, dfz) which is obtained from the "F-Tables" that are based on the respective degrees of freedom for the estimates of s_a, and s_b. See Appendix at the end of the chapter for the F tabulated values. If F is greater than F_{table}, then s_a^2 can be considered to be greater than s_b^2, at the chosen level of confidence. If F is less than F_{table},, then there is no reason to believe that s_a is greater than s_b at the chosen level of confidence.

To use the F test, we assume that variances of the two populations are equal. This hypothesis is called mull hypothesis:

$$H_0 : \sigma_A = \sigma_B$$

And the alternative hypothesis, that the two variance are different

$$H_1 : \sigma_A \neq \sigma_B$$

If the value of F calculated is inferior to that tabulated we accept the null hypothesis that the two variances are equal. Otherwise, we accep the alternative hypothesis.

Example

Two analysts determined the copper content in a beverage six times. Do these analysts differ in the precision of their determination at $\alpha = 0.05$? We apply the F-test

$$F = \frac{s_1^2}{s_2^2}$$

Where s_1 is the higher standard deviation.

Data:

Analyst 1	Analyst 2
$s_1 = 1.80$	$s_2 = 1.20$
$n_1 = 6$	$n_2 = 6$
$d\nu = 5$	$d\nu = 5$

Where n1 and n2 are the number of measurement and δv the degree of freedom.

$$F = \frac{3.24}{1.44} = 2.25$$

Our experimental value is 2.25 while the tabulated value for 5 degree of freedom using a two-tailed test (p = 0.05) is:

$$F_t = 5.05$$

Considering that 5.05 > 2.25 we accept the null hypothesis that the two standard deviation are not different.

If we want to know if $s_1 > s_2$ we must use the on-tail test. In this case the value of F (p=0.025) is 5.05. We conclude that there is no reason to retain that s_1 is higher than s_2.

4.12 Outliers

4.12.1 Outliers-Dixon Test

In a set of observations one can encounter an unusually large or small value or measurement that does not belong with the other values or measurements. This type of value is usually referred to as an outlier. There are many possible reasons for an outlier, such as blunder, malfunction of the method, calculation error, error in transcribing data, or some unusual loss or contamination. When a result appears to be an outlier, it can be subjected to a test to see whether it can reasonably be rejected. The usual purpose of the test is to identify the need for closer control of the process, or to eliminate the outlier prior to evaluation of the data. This data preprocessing should be done with caution, since it is not advisable to presume that test data are defective without first trying to identify the cause of the problem. If an outlier is discarded, that fact should be reported.

There are several statistical tests in common use for identifying outliers for normal populations. Two such tests are the Grubbs Test and the Dixon Test. These tests are not difficult to follow, and they are described as follow.

Dixon test: in order to apply this test, it is necessary:

1. To arrange values in increasing value
2. To decide which is the suspected value (the higher or the lower)
3. To select the risk (i.e. 95%, 99% and so on)
4. To calculate the R value on the basis of the number of data and the suspected value (lower or higher)
5. To compare the R value with that of the Dixon table in the Appendix at the end of the chapter.

n	Ratio	Suspected higher x_n	Suspected lower x_1
$3 \leq N \leq 7$	R	$X_N - X_{N-1})/(X_N - X_1)$	$(X_2 - X_1)/(X_N - X_1)$
$3 \leq N \leq 10$	R_1	$(X_N - X_{N-1})/(X_N - X_2)$	$(X_2 - X_1)/(X_{N-1} - X_1)$
$11 \leq N \leq 13$	R_2	$(X_N - X_{N-2})/(X_N - X_2)$	$(X_3 - X_1)/(X_{N-1} - X_1)$
$14 \leq N \leq 25$	R_3	$(X_N - X_{N-2})/(X_N - X_3)$	$(X_3 - X_1)/(X_{N-2} - X_1)$

Example

Consider the following data:

$$0.42-0.35-0.38-0.39-0.40-0.47-0.40-0.43-0.37$$

Arrange the values in increasing order:

$$0.35-0.37-0.38-0.39-0.40-0.40-0.42-0.43-0.47$$

We decide that 0.47 is the suspected value. The R1 value calculated following the schema depicted above is:

$$R_1 = \frac{0.47-0.43}{0.47-0.37} = 0.364$$

The value of R_1 is compared with the critical value in the Dixon Table in the appendix at the end of the chapter, that in this case is $R_1 = 0512$.

Conclusions: Since the critical value is not exceeded, the suspected value is not considered an outlier and cannot be discarded.

4.12.2 Grubbs Test

- If the values really were all sampled from a Gaussian distribution, what is the chance that you would find one value as far from the others as you observed? If this probability is small, then you will conclude that the outlier is not from the same distribution as the other values. Assuming you answered no to all three questions above, you have justification to exclude it from your analyses. Statisticians have devised several methods for detecting outliers.

Grubbs' test is one of the most popular ways to define outliers, and is quite easy to understand. The first step is to quantify how far the outlier is from the others. Calculate the ratio Z as the difference between the outlier and the mean divided by the SD. If Z is large, the value is far from the others. Note that you calculate the mean and SD from all values, including the outlier.

Procedure for using the Grubbs test:

1. Rank the data
2. Decide if the largest or the smallest data is the outlier
3. Estimate the standard deviation of all data (s)
4. Calculate G value as follow:

$$G = \frac{x_{average} - x_1}{s} \quad or \quad \frac{x_n - x_{average}}{s}$$

5. Select the risk for false rejection (see table below)
6. Compare the G obtained with that of the table considering the number of data (n) and the risk selected.

If G is larger than the value tabulated, rejection can be made.

Example
4.18; 4.19; 4.19; 4.20; 4.21; 4.22; 4.22; 4.24; 4.25; 4.29
Average = 42.19/10 = 4.219 Standard deviation (s) = 0.0318

$$G = \frac{4.29?4.219}{0.0318} = 2.23$$

G value in the table in Appendix 1 at 5% risk is G = 2.176
Conclusion: There is reason to reject 4.29

4.13 Cochran Test for Extreme Value of Variance (Outlier Variance)

This test maybe used to decide whether a variance is larger with respect a group of comparable variance. It can be applied for instance to compare variance obtained by a group of laboratories participating to a proficiency testing. It is based on the table reported below. The variances are summed $\sum$ si2 and the ratio of the largest variance sl2 to the summed $\sum$ si2 is calculated. If this ratio exceed the value of the Cochran reported in the table is will considered an outlier with 95% confidence. All variance should have the same number of degree of freedom.

Example
Variances: 2.25; 4.41; 2.25; 4.0; 3.24; n = 3
$\sum$ = 16.15 Cochran ratio 4.41/16.15 = 0.273
From the Cochran table in the appendix at the end of the chapter, the value for 5 variances obtained with 3 replicates is 0.6838 that is larger than the calculated value. Conclusion: The value 4.41 is not an extreme value and should be retained.

Level of significance 95%					
N. of variance compared	Number of replicate value				
	2	3	4	5	6
2	0.9985	0.9750	0.9392	0.9057	0.8772
3	0.9669	0.8709	0.7977	0.7457	0.7071
4	0.9065	0.7679	0.6841	0.6287	0.5895
5	0.8412	0.6838	0.5981	0.5441	0.5065
6	0.7808	0.6161	0.5321	0.4803	0.4447
7	0.7271	0.5612	0.4800	0.4307	0.3974
8	0.6798	0.5157	0.4377	0.3910	0.3595
9	0.6385	0.4775	0.4027	0.3584	0.3286
10	0.6020	0.4450	0.3733	0.3311	0.3029

4.14 Combining (Pooling) Estimates of Standard Deviations

It is possible to pool (combine) several estimates of standard deviation to obtain a better estimate of standard deviation (i.e., with more degrees of freedom) using the equation below, where v_i is the number of degrees of freedom of the ith estimate of s, s1 and s2 the standard deviation and dν1 and dν2 the degree of freedom

$$Spooled = s_p = \sqrt{\frac{s_1^2 x d_{v1} + s_2^2 x d_{v2} + \ldots}{d_{v1} + d_{v2} + \ldots}}$$

Example
The standard deviation for a given measurement have been estimated four times in different occasions. Calculate the pooled standard deviation. The results are reported in the following table:

series	s	n	dν	s^2	dν . s^2
1	1.62	9	8	2.624	20.99
2	1.80	6	5	3.240	16.20
3	2.15	4	3	4.623	13.87
4	1.50	12	11	2.250	24.75

On the basis of the relation above reported the pooled standard deviation is:

$$s_p = \sqrt{\frac{20.99 + 16.20 + 13.87 + 24.75}{8 + 5 + 3 + 11}} = 1.68$$

4.15 Precision Calculations

As we have seen previously the standard deviation provides an estimate of the variability of the data. It can be calculated in different ways depending on the analytic situation. Below are some illustrative examples.

Way 1
Calculation of the standard deviation of a series of analyses performed several times on a single sample. The well-known following relation will be used:

$$s = \sqrt{\frac{\Sigma(x_i - m)^2}{n - 1}}$$

where

s = standard deviation
n = number of analyses
$d\nu = n - 1$
m = average of data

We obtain the following data:

x_i	$x_i - m$	$(x_i - m)^2$
4.1	−0.06	0.0360
4.3	0.14	0.0196
3.8	−0.36	0.1296
4.4	0.24	0.0576
4.2	0.04	0.0016
x = 4.16		Σ 0.244

And the standard deviation is:

$$s = \sqrt{\frac{0.244}{4}} = 0.25$$

And the relative standard deviation RSD

$$RSD = \frac{0.247}{4.16} \times 100 = 5.94\%$$

Way 2
The standard deviation can be calculated also by utilizing the Range and the following relation:

$$s = \frac{x_g - x_p}{d_2}$$

Where

x_g = the higher value of the series of data
x_p = the lower value of the series of data
d_2 = tabulated value see Table xxx

In the above example we obtain the following results

$$s = \frac{4.4 - 3.8}{2.24} = \frac{0.6}{2.24} = 0.26$$

Way 3

It should be reminded that the standard deviation might be estimated from the differences of several set of duplicate measurements on the same sample in different occasions or on different sample with similar composition.

The equation for these calculations is:

$$s = \sqrt{\frac{\sum d^2}{2k}} \tag{4.1}$$

Where k is the number of duplicate measurements and d the difference between duplicates.

Example 1

A sample has been analyzed in duplicate in seven different occasions. The results are reported in the following table:

Experiment	x_1	x_2	d	d^2
1	11.5	11.2	+0.3	0.09
2	11.8	11.7	+0.1	0.01
3	11.2	11.9	−0.7	0.49
4	11.7	11.1	+0.6	0.36
5	11.9	11.2	+0.7	0.49
6	11.4	11.7	−0.3	0.09
7	11.3	11.5	−0.2	0.04
				Σ1.48

By using the above Eq. 4.1 the standard deviation (s) can be calculated.

$$s = \sqrt{\frac{1.48}{2 \times 7}}$$

Now consider the situation where the laboratory, operating under repeatability conditions, obtain two results (x_1, x_2) K is equal to 1 and the Eq. 4.1 became:

$$s\sqrt{2} = x_1 - x_2 = \mathbf{d}$$

By using the coverage factor 2 the difference between the two results should not exceed

$$2\sqrt{2}s_r = 2.83 s_r = r$$

where r is the repeatability limit. (see point 4.18).

Knowing the repeatability limit the expected standard deviation of the method can be calculated.

Example 2
The same relationship can be used for the analysis made in duplicate on different sample with similar composition.

In the following table are reported the results obtained analysing six type of different cheeses for their fat content.

Experiment	x_1	x_2	d	d^2
1	30.56	30.62	−0.06	0.0036
2	26.70	26.60	+0.10	0.0100
3	24.90	24.82	+0.08	0.0064
4	37.10	36.90	+0.20	0.0400
5	30.45	30.40	+0.05	0.002
6	30.62	31.73	−0.11	0.0121
7				Σ0.0746

The standard deviation, calculated with the usual relationship will be:

$$s = \sqrt{\frac{0.0746}{2 \times 6}}$$

4.16 Averages

4.16.1 Comparison of Means: The T-Test

The t-test can be used to determine whether there is a statistically significant difference between the means of two sets of values at a given probability level. The statistic t is given by the following equation:

$$t = \frac{m_1 - m_2}{s_p}$$

The two means are m_1 and m_2, and s_p is the pooled standard deviation (see above).

The value obtained for tis then compared with a tabulated value of that corresponds to the chosen level of significance. This comparison permits two alternatives: the first decides that there is a real difference between the two averages; the second decides that there is insufficient evidence to justify a claim that the samples differ.

This type of testing procedure, is known as significance testing and it involves the formulation of the "null-hypothesis." The t-test can also be used to test whether a sample mean is significantly different from an expected or reference value. In this case, the reference value is R_{ef} and we use the following relationship.

$$t = \frac{m_1 - R_{ef}}{\frac{s}{\sqrt{n}}}$$

Where s is the standard deviation and the degrees of freedom for t are n − 1

Does the average agree with the theoretical or an accepted value?

Worked Example

A sample of foodstuff has a known content of fat of 9.55%.

By analyzing the sample we obtain the following values:

Fat %	9.17	9.09	9.14	9.10	9.13	9.27

We calculate the average and the standard deviation:

$$\text{m} = 9.15 \quad \text{s} = 0.065 \quad d\nu = 5$$

WE formulate the null-hypothesis and the alternative hypothesis:

$$H_0 : \mu_1 = 9.55$$

$$H_1 : \mu_1 \neq 9.55$$

and the level of significance of 95% ($p = 0.05$).

On the basis of the following relation we calculate the t value

$$t = \frac{m_1 - R_{ef}}{\frac{s}{\sqrt{n}}}$$

where R_{ef} is the designed value.

$$t = \frac{9.51 - 9.55}{\frac{0.065}{\sqrt{6}}} = \frac{0.40}{0.0267} = 14.8$$

This t value is compared with that tabulated (see table t in Appendix at the end of the chapter) for the probability level of 95%:

$$t_{abulated} = 2.57$$

Conclusion: Since our calculated value is higher than that tabulated (14.98 > 2.57) we reject the null hypothesis and we conclude that there is a difference between the worked average and the designed for a probability level of 95%.

4.17 Comparing Two Averages by Using the T-Test

In the following three examples given the analyst should apply the right t-test and interpret the results. Results can be checked using the built in functions of a spreadsheet like Excel or a statistical software package.

Example 1

A chemist is asked to validate a new faster method for the determination of nitrate in samples of milk powder. The reference sample analysed by the old method shows a mean of 12.7 ppm (m_A). For the new method the mean is 13.5 ppm (m_B), based on 10 results with a standard deviation (s) of 0.85 ppm. Is the new method equivalent to the old one?

Answer

First of all, we should state the null and the alternative hypotheses.

The null hypothesis is: H_0: $m_a = m_b$

The alternative hypothesis is: H_1: $m_a \neq m_b$

Confidence level 95%

Degrees of freedom: 9

In this case we wish to know if the means are different so the two-sided critical value should be used.

The formula to be used is:

$$t = \frac{m_A - m_B}{\frac{s}{\sqrt{n}}} = \frac{13.5 - 12.7}{\frac{0.85}{10}} = 2.976$$

From t- tables, $t_{crit} = 2.26$ at the 95% confidence level for 9 degrees of freedom.

Since $t_{calc} > t_{crit}$ we can reject the null hypothesis and conclude that we are 95% certain that there is a significant difference between the new and old methods.

Example 2

Two batches of a product are analysed for the iron content with an AAS method which has a standard deviation of 0.025 ppm. Based on the data shown in following table, do the two batches differ significantly in the iron content? Use a 95% level of significance and a 2-tailed test.

Batch	Method 1 (ppm)	Method 2 (ppm)
1	9.1	4.2
2	4.0	4.5
3	5.2	7.2
4		6.8
5		4.3
Mean	6.10(m_B)	5.40 (m_A)
Standard dev. (s)	$s_a = 2.667$	$s_b = 1.47$

Answer to Example 2

First of all we should verify if the two standard deviations are significantly different, by applying an F-test (see Note 1).

$$F = \frac{s_a^2}{s_b^2} = \frac{2.667^2}{1.471^2} = 3.287 \quad F_{\text{tab}2,4} = 10.6$$

Since 10.6 > 3.287 at a confidence level of 95% there no difference between the two variances then we can combine (or pool) the standard deviation using the following relation:

$$s_p = \sqrt{\frac{4 \times 1.471^2 + 2 \times 2.667^2}{(5-1)+(3-1)}} = 1.953$$

And

$$d\nu = \text{n}_1 + \text{n}_2 - 2 = 6$$

At this point we can formulate the null and the alternative hypotheses:

The null hypothesis is: H_0: $m_a = m_b$
The alternative hypothesis is: H_1: $m_a \neq m_b$
Confidence level 95%.

In this case we wish to know if the means are different so the two-sided critical value should be used.

$$t = \frac{m_A - m_b}{s_p\sqrt{\frac{1}{n_a} + \frac{1}{n_b}}} = \frac{5.40 - 6.1}{1.953\sqrt{\frac{1}{3} + \frac{1}{5}}} = \frac{0.70}{1.426} \cong 0.490$$

The tabulated value of t_{crit} (with v = 6 degrees of freedom, at the 95% confidence limit) is 2.447.

Since $t_{calc} > t_{crit}$ we can reject the null hypothesis and conclude that we are 95% certain that there is a significant difference between the two methods.

Example 3

Two methods are available for determining the concentration of vitamins in foodstuffs. In order to compare the methods several different sample matrices are investigated using the same technique. Each matrix type is divided into two aliquots and the vitamin content is measured by the two methods. Results are reported on the following table.

Sample	Method A	Method B	Difference	Difference^2
1	2.52	3.17	−0.65	0.423
2	3.13	5.00	−1.87	3.497
3	4.33	4.03	0.30	0.09
4	2.25	2.38	−0.13	0.017
5	2.79	3.68	−0.89	0.792
6	3.04	2.94	0.10	0.01
7	2.19	2.83	−0.64	0.410
8	2.16	2.18	−0.02	0.0004
			$\bar{d} = -0.475$ $(\Sigma d)^2 = 3.8$	$\Sigma d^2 = 5.239$

Do the two methods give different average vitamin contents?

Answer to Example 3

First of all, we state the null and the alternative hypotheses.

The null hypothesis is H_0: d = 0

The alternative hypothesis is H_1: d ≠ 0

The test is a two tailed test as we are interested in both d < 0 and d > 0.

The mean of differences is $\bar{d} = |-0.475|$

The sample standard deviation of the paired differences is $s_d = 0.700$ calculated as follows:

$$s_d = \sqrt{\frac{\Sigma d^2 - \frac{(\Sigma d)^2}{n}}{n-1}}$$

The following equation is used for the t test on paired data:

$$t = \frac{\bar{d}}{s_d}\sqrt{n}$$

where $\bar{d}$ is the average difference in the pairs of values. The absolute value is generally used.

The tabulated value of t_{crit} (with v = 7 degrees of freedom, at the 95% confidence limit) is 2.365.

Since the calculated value is less than the critical value Ho cannot be rejected. It follows that there is no difference between the two techniques.

4.18 The Repeatability Limit (r)

The repeatability limit is the limit that should not be exceeded by the absolute difference between two test results obtained under repeatability conditions. It is defined as:

$$r = 2.83\,s_r$$

Where s_r is the repeatability standard deviation and r the repeatability limit.

The term r, named repeatability limit, may be interpreted as the amount within which two determinations should agree with each other within a laboratory 95% of the time.

Estimates of s_r can be obtained only from a planned, organised method-performance study. Estimate of s_r can be also obtained from routine work within a laboratory by use of control charts.

Another question: Why the repeatability limit is 2.83 $s_{r.}$

It should be reminded that the standard deviation may be estimated from the differences of several set of duplicate measurements on the same sample in different occasions or on different sample with similar composition.

The Eq. 4.1 for these calculations is used:

$$s = \sqrt{\frac{\Sigma d^2}{2k}}$$

Where k is the number of duplicate measurements and d the difference between duplicates.

If K = 1, we have

$$s = \sqrt{\frac{\Sigma d^2}{2}}$$

Consequently,

$$s\sqrt{2} = d \;\; or \;\; s \times 1.41 = d$$

And multiplying for the coverage factor 2, we have that:

$$s \times 1.41 \times 2 = d = s \times 2.83$$

When test results do not meet the condition of repeatability the flow diagram reported below should be followed

r = repeatability limit = 2.83 s_r
μ = final quoted result

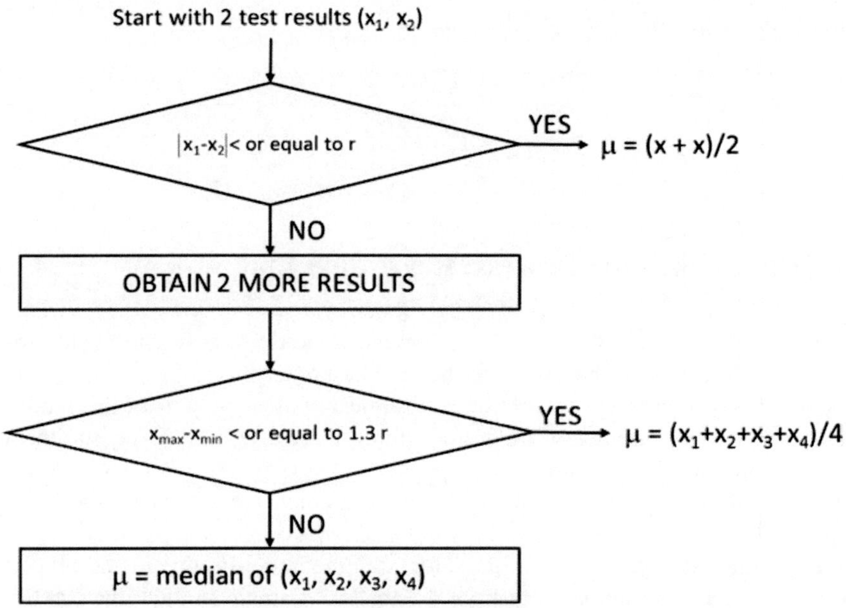

An example of application of the repeatability limit is reported below. It refers to the statistical analysis of Collaborative Study Results.

Example

The standard deviation and the repeatability limit are calculated from the results obtained in duplicate by 11 laboratories by analysing the same sample.

Lab N°	a	b	y	w	w^2
1	7.8	7.6	7.70	0.2	0.04
2	8.8	7.2	8.00	1.6	2.56
3	7.7	9.7	8.70	2.0	4.00
4	8.7	8.5	8.60	0.2	0.04
Z < 5	7.8	7.4	7.60	0.4	0.16
6	9.0	9.2	9.10	0.2	0.04
7	8.1	8.4	8.25	0.3	0.16
8	9.9	9.7	9.80	0.2	0.04
9	7.6	7.9	7.75	0.3	0.09
10	8.4	8.3	8.35	0.1	0.01
11	7.8	8.4	8.10	0.6	0.36
			$\Sigma y = 91.95$		$\Sigma w^2 = 7.43$
			$M = \Sigma y/n = 8.36$		

w = difference between duplicate and y = average

The results are reported in the following table.

s_r^2	$\Sigma w^2/2n$	7.43/22	0.3377
s_r	$\sqrt{s_r^2}$	$\sqrt{0.3377}$	0.58
r	$2.83 \times s_r$		1.64

4.19 The Calibration Process: Regression Line

Calibration is the process of establishing how the response of a measurement process varies with respect to the parameter being measured.

The usual way to perform calibration is to subject known amounts of the parameter (e.g. using a measurement standard or reference material) to the measurement process and monitor the measurement response.

Goal of calibration: Ability to calculate a (measurement) result in a secure (safe) working range

Fitting a straight line to a plot of data, for example in constructing a calibration curve, is a frequent and important task for the analyst. A single straight line that lies closest to all of the points on a scatter plot is known as a regression line. The line can be used for making certain predictions when the following values are known: the slope of the line and the point of intercept (intersection) with the y-axis.

The equation for least square regression lines is:

$$y = a + bx \tag{4.1}$$

It is found from data $y_1, y_2, y_3 \ldots\ldots, \ldots y_n$ associated respectively with $x_1, x_2, x_3, \ldots x_n$. If only the observations of y are affected by significant error, the least-squares-estimates for the parameters a and b can be obtained by the following formulas:

$$incercept = a = y_{av} - bx_{av} \tag{4.2}$$

$$slope = b = \frac{(\Sigma x_i - x_{av})(y_i - y_{av})}{\Sigma(x_i - x_{av})^2} \tag{4.3}$$

The sum of the deviations squared will be a minimum, that is, any other straight line that can be drawn would have a greater sum of squared deviations.

A useful statistic, which estimates the random errors in the y-direction, is

$$s_{x/y} = \sqrt{\frac{\Sigma(y_i - y_{average})^2}{n-2}} \tag{4.4}$$

This will allow calculating the standard deviation of the slope and the intercept by the following relationship:

$$\textit{standard deviation of the slope} = s_b = \frac{s_{x/y}}{\sqrt{\sum\left(x_i - x_{average}\right)^2}} \tag{4.5}$$

$$\textit{standard deviation of the intercept} = s_a = s_{x/y}\sqrt{\frac{\sum x_i^2}{n\sqrt{\sum\left(x_i - x_{average}\right)^2}}} \tag{4.6}$$

Calculation can be done with a specific software (recommendable) or manually by using the example below.

Worked Example
Considering the following data,

Concentration, x	0	1	2	3	4	5	6
Instrument response, y	0.7	1.7	3.0	4.2	5.8	7.0	8.2

the regression line, the standard deviation and confidence limit of the slope and intercept of the regression line can be calculated by using Excel or any other available software. The following graph is obtained by using Excel:

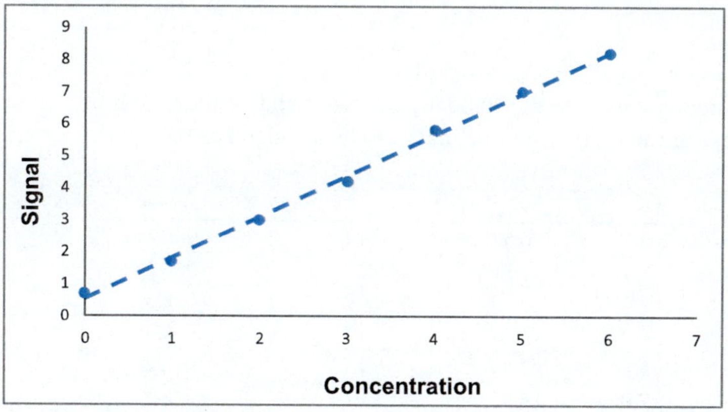

To better organize the calculation, the following table can be set.

x_i	y_i	$x_i - x_{av}$	$(x_i - x_{av})^2$	$y_i - y_{av}$	$(y_i - y_{av})^2$	$(x_i - x_{av})x(y_i - y_{av})$
0	0.7	−3	9	−3.67	13.47	11.01
1	1.7	−2	4	−2.67	7.13	5.34
2	3.0	−1	1	−1.37	1.88	1.37
3	4.2	0	0	−0.17	0.03	0

	x_i	y_i	$x_i - x_{av}$	$(x_i - x_{av})^2$	$y_i - y_{av}$	$(y_i - y_{av})^2$	$(x_i - x_{av})x(y_i - y_{av})$
	4	5.8	1	1	1.43	2.05	1.43
	5	7.0	2	4	2.63	6.92	5.26
	6	8.2	3	9	3.83	14.67	11.49
Sum	21	30.6	0	28	**0**	46.15	**35.9**
Average	3	4.37					

Using these totals and the equation for the correlation, we can calculate the correlation coefficient that is used to measure how strong is the relationship between two variables (see the tip at the end of the paragraph):

$$r = \frac{\Sigma(x_i - x_{av})x(y_i - y_{av})}{\sqrt{(x_i - x_{av})^2 (y_i - y_{av})^2}} \tag{4.7}$$

For the example:

$$r = \frac{35.9}{\sqrt{28 \times 46.15}} = 0.9989$$

The slope:

$$b = \frac{35.9}{28} = 1.282$$

The intercept: $a = 4.37 - (1.282 \times 3) = 0.524$

Finally, to manually calculate the standard deviation and confidence limit of the slope and intercept of the regression line, data can be arranged as follow:

x_i	x_i^2	y_i	$y_{i\,regline}$	$y_i - y_{i\,regline}$	$(y_i - y_{i\,regline})^2$
0	0	0.7	0.52	0.18	0.032
1	1	1.7	1.81	0.11	0.012
2	4	3.0	3.09	0.09	0.008
3	9	4.2	4.37	0.17	0.029
4	16	5.8	5.65	0.15	0.022
5	25	7.0	6.93	0.07	0.005
6	36	8.2	8.22	0.02	0.000
Sum	91				0.108

Where $y_{i\,regline}$ are the values calculated from the regression line.

Now we can calculate the $s_{x/y}$ statistics:

$$= s_{x/y} = \sqrt{\frac{0.108}{5}} = \sqrt{0.0216} = 0.146$$

From the above equations, it can be shown that standard deviation of the slope is

$$s_b = \frac{0.146}{\sqrt{28}} = 0.027$$

And considering that the t value for (n – 2) = 5 at the 95% confidence level is 2.57, the confidence limit for the slope will be:

$$\mathrm{b} = 1.282 \pm (2.57 \times 0.027) = 1.282 \times 0.069$$

while for the intercept the standard deviation s_a will be:

$$s_a = 0.146\sqrt{\frac{91}{7 \times 28}} = 0.099$$

and the confidence limit a = 0.524 ± (2.57 × 0.099) = 0.524 ± 0.25

Once we calculated the regression line we can calculate any x-value corresponding to any measured signal (y–value), and to evaluate the error of the x-value we can use the following relationship whose value can be excerpted by the above calculations:

$$s_x = \frac{s_{y/x}}{b}\sqrt{\frac{1}{n} + \frac{1}{m} + \frac{(y - y_{av})}{b^2 \sum (x - x_{av})^2}} \tag{4.7}$$

For instance, for an y-value of 4.5, the corresponding x-value from the regression line is 3.10. The corresponding confidence limits (t = 2.57) is 3.10 ± 0.115. Many software, Excel included, can easily do those calculations.

4.20 Weighted Regression Line

Sometimes experimental data indicate that errors increase as the analyte mass concentration increases, indicating heteroscedastic data. The homoscedasticity assumption must be tested in any linear regression analysis. It can be performed by plotting residuals versus concentration and by applying an F-test using the variance obtained at the lowest and highest concentration level of the working range.

In case of heteroscedastic data, the regression line is calculated to give additional weight to those points where errors are the smallest. Therefore, the mass concentration (C) is calculated using the weighted regression method. Examples of calculation can be found in the recommended book at the beginning of the chapter.

4.20.1 *TIP*

4.20.1.1 The Correlation Coefficient

Linearity is usually measured by the square of the correlation coefficient for the calibration curve. Correlation coefficient formulas are used to find how strong a relationship is between data.

- A correlation coefficient of 1 means that for every positive increase in one variable, there is a positive increase of a fixed proportion in the other. It indicates strong positive relationship.
- A correlation coefficient of −1 means that for every positive increase in one variable, there is a negative decrease of a fixed proportion in the other. It indicates a strong negative relationship.
- Zero means that for every increase, there is not a positive or negative increase. It indicates no relationship at all.

Cautions
The correlation coefficient is a measure of the extent to which the regression model explains the variation in y. However, most straight-line calibration curves have a correlation coefficient very close to +1, typically 0.99 or better, showing that the linear regression model is appropriate even though the plot of the data looks to be curvilinear.

4.21 Method of Standard Addition (MOSA)

The regression line can be used also when the laboratory wants to perform the standard addition method. This method is highly recommended to evaluate/compensate for matrix interferences in the analysis and for complex matrix samples. It is applied more often in atomic absorption spectroscopy and in electrochemical methods of analyses. The method consists in the preparation of solutions of equal amount of sample to which know amount of standards except one, are added known concentration of the analyte before completing volume with the solvent. The instrumental signals of the standard solutions are recorded and a calibration curve is created. The sample concentration is calculated by extrapolation to zero of the linear equation.

The following exercise will clarify the procedure.

The vitamin C in a sample of orangeade was determined electrochemically by the method of standard additions. Five solutions of 5 ml sample were poured in a 10 ml flask. The first flask containing the sample was simply diluted to the 10 ml with water. To all the other increasing amounts, as reported in the table below, of standard solution were added before dilution to the 10 ml mark. The following results were obtained.

A graph of signal versus concentration in milligrams per liter can be prepared as reported in Fig. 4.3.

Vitamin C added mg/L (spikes)	0	15	30	45	60
Signal mV	0.3	0.42	0.51	0.60	0.72

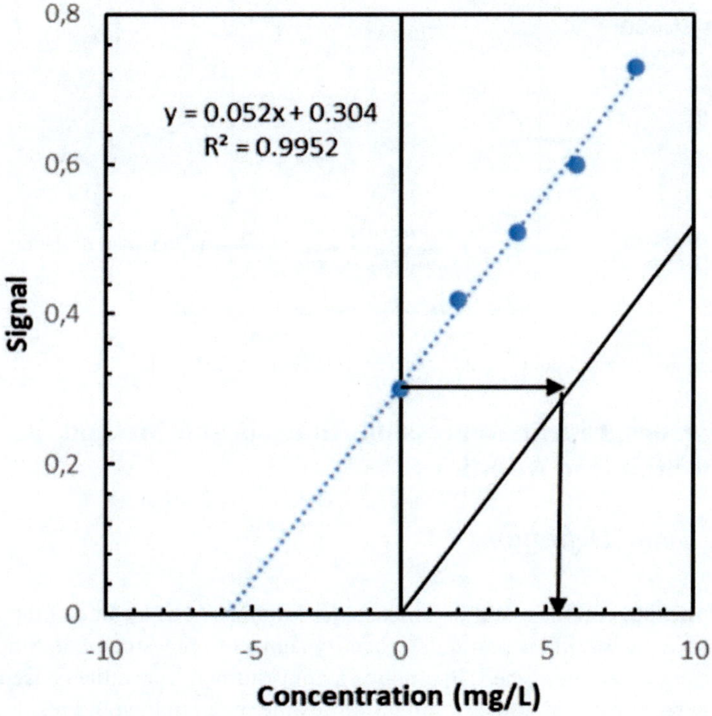

Fig. 4.3 MOSA plot

Calculation by using Excel give the following regression line:

$$y = 0.052x + 0.304$$

By setting y = 0 in the Eq. 4.5, the ratio 0.304/0.052 gives the amount of analyte (5.85 mg/L) in the test sample that should be multiplied by the dilution factor (2 in this example).

Once we calculated the regression line we can calculate any x-value corresponding to any measured signal (y–value) and to evaluate the error of the x-value we can use the below formula that is similar to the one used for the regression line constructed with standard solutions:

$$s_x = \frac{s_{y/x}}{b}\sqrt{\frac{1}{n} + \frac{y_{av}}{b^2 \sum (x - x_{av})^2}} \tag{4.8}$$

The confidence limit can be calculated by using the above relationship.

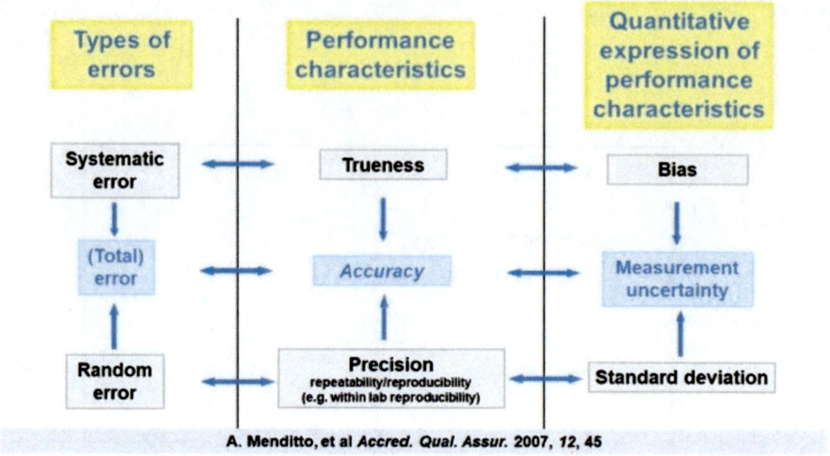

Fig. 4.4 Scheme from Menditto et al. in Accred. Qual. A ssur. 207,12,45

4.22 Errors, Linear Regression Analysis and Method of Standard Additions

4.22.1 Some Definitions

error (of measurement) result of a measurement minus a true value of the measurand where measurand is particular quantity subject to measurement. Since a true value cannot be determined, in practice a conventional true value is used.

random error: result of a measurement minus the mean that would result from an infinite number of measurements of the same measurand carried out under repeatability condition. Random error is equal to error minus systematic error.

systematic error: mean that would result from an infinite number of measurements of the same measurand carried out under repeatability conditions minus a true value of the measurand. Systematic error is equal to error minus random error.

Random error corresponds to imprecision, and bias to inaccuracy. Here is a diagram that will attempt to differentiate between imprecision and inaccuracy (Fig. 4.4). The error of the result of a measurement may often be considered as arising from a number of random and systematic effects that contribute individual components of error to the error of the result.

Bias in an analytical method is particularly difficult to detect and the best way to estimate the bias of an analytical method is by analyzing a Standard Reference Material if available or a synthetic standard material that closely approximate the composition of the samples to be analyzed. If standard samples are not available, a second independent and reliable analytical method can be used in parallel. However, the independent method should differ as much as possible from the one under study and based on different analytical technique, i.e. spectrophotometric vs electrochemical.

4.22.2 Errors in Chemical Analysis

Chemical analyses are mainly affected by two types of errors:

1. Random (or indeterminate) error, causes data to be scattered more or less symmetrically around a mean value.
2. Systematic (or determinate) error, lead to bias in measurement results and causes the mean of a data set to differ from the accepted value. It may be either constant or proportional.

Systematic errors in turn can be divided into and incorrigible and corrigible errors. The former due to the calibration process or to direct interference, can be detected by a direct examination of the experimental conditions or by a prior knowledge of the history of the sample but cannot be diagnosed by statistical means and therefore cannot be corrected.

The latter, which we can distinguish in constant or proportional, can be diagnosed by means of statistical tests. Their magnitude can be determined and therefore be corrected. An example of a constant error can be that of non-specific absorptions in AAS.

4.22.3 Constant Error

The magnitude of the constant errors is essentially the same as the size of the quantity measured is varied. So the absolute error is constant with sample size, while the relative error varies when the sample size is changed. As the size of a measurement increases, the effect of a constant error decreases. Thus, constant errors can often be detected by varying the sample size.

4.22.4 Proportional Errors

Proportional errors decrease or increase in proportion to the size of the sample. A common cause of proportional errors is the presence of interfering contaminants in the sample.

A proportional error, on the other hand, results from a significant change in response per unit of analyte concentration attributable to a parameter of the system or to the specific analytical procedure. They are independent of the amount of sample used in the analysis for example caused by the presence of interfering substances.

4.23 The Youden Approach to Constant and Proportional Errors

A blank contains the reagents and solvents used in a determination, but no the analyte.

Often, many of the sample constituents are added to simulate the analyte environment, which is called the sample matrix.

In a blank determination, all steps of the analysis are performed on the blank material. The results are then applied as a correction to the sample measurements.

Blank determinations reveal errors due to interfering contaminants from the reagents and vessels employed in the analysis.

The direct determination of the constant error consists in the measurement of the blank. There are two types of blank: the blank of the system which is obtained by analyzing a solution that does not contain the sample subjected to the entire method of analysis; the method blank which is the response given by a sample that does not contain the analyte (placebo) subjected to the entire method of analysis. The latter allows us to correct the constant error in analytical calculations.

The blank's responses are algebraically subtracted from the responses due to the sample (MB) and the standards (SB). The blank is therefore a valid method to verify the influence on the analytical signal by external agents such as reagents, analytical instrumentation and the external environment. Wineforder says that "the blank is a solution containing all that is present in the sample except the analyte." The MB depends, however, on the quality of the placebo being similar to the sample without the analyte of interest. The true blank however is that which is determined using the sample in which both the analyte and the sample are present.

It is evident that the blank, used as a measure of the constant error, must be calculated when the analyte and the matrix are present at the same time.

Youden solved the problem by using linear regression of the data obtained by analyzing different sample takes. In practice there is a calibration line in which the signals are shown on the y axis and the corresponding sample takes on the x axis. The consequent linear regression gives us a slope and an intercept that represents the total blank (TBY) that is the constant error of the method. This is the only value to use in correcting the data as the linear regression represents the standard curve of the analyte in the presence of the matrix.

As was shown in the previous section, the conventional analytical procedure that requires a calibration curve based on the analysis of standards (straight line) is based, as seen before, on the following equation:

$$Y = m_s C + SB \tag{4.1}$$

from where concentration (C) can be calculated:

$$C = \frac{Y - SB}{m_s} \tag{4.2}$$

where m_s = slope of the standard curve, SB = the intercept (standard blank), and Y the response.

The term SB and his confidence interval, at some confidence level, we saw that can be used to calculate the detection limit and can be considered as a calibration constant. If zero is included in his confidence interval can be said that the intercept is not significantly different from zero and that there is not bias error in the calibration. However, if the intercept value is statistically significant from zero, bias error is present and causes for it should be investigated.

It is important to note that a similar calibration regression line has been used for the method of standard addition (MOSA). The standard addition line is an in situ normalization procedure for the proportional error. This normalization is due to the change of the slope of the MOSA curve with respect to the slope of the curve of the standard solutions. Although the MOSA can compensate for proportional determinate errors, it cannot correct for a constant determinate error. To correct for a constant method error, a blank must account for signal s from any reagents and solvents used in the analysis, as well as any bias resulting from interaction between the analyte and the sample matrix. Unfortunately, neither a calibration blank not a reagent blank can compensate from an interaction between the anlyte and the sample's matrix. To be effective, the blank must include both the sample's matrix and the anlyte and, consequently, must be determined using the sample itself. Youden approach is to measure the signal from different sample size, and to determine the regression line. The resulting y-intercept gives the signal in the absence of the sample, and it is known as the total Youden blank.

$$Y = m_s C + TYB \tag{4.3}$$

And

$$C = \frac{Y - TYB}{m_s} \tag{4.4}$$

The two curves 1 and 3 represent the responses due to the standards but the first in pure solvent and the second in the presence of the sample matrix. If the two lines are parallel and therefore have the same slope, there is not proportional error; otherwise, the ratio between the two slopes can be used to calculate the proportional error.

$$P = \frac{m_M}{m_s}$$

This ratio P should be equal to unity unless proportional error is present. In the first case the value of P is simply a calibration constant. If the slope ratio confidence interval of P does not contain the unity a proportional systematic error exists and search for the causes should be investigated.

In practice, for preventing possible matrix effects, the Method of Standard Additions (MOSA) should be applied in conjunction with the conventional external calibration. First, a normal calibration curve of standards versus concentrations should be prepared to determine the value of the slope, m_s. Next, a standard additions calibration curve by plotting the data as shown in Fig. 4.3, should be prepared. This slope provides an independent result of m_M. If there is no significant difference between the two values of the slopes, then the difference between the sample's matrix and that of the external standards can be ignored. On the contrary, using a normal calibration curve, a proportional determinate error will be introduced.

However, as pointed out by Cardone et al. (1,2,3,4) a prior step in the application of MOSA, is the the determination of the Total Youden Blank (TYB). In fact, the analytical blank (YB) cannot be subtracted from the analytical signal, because it comes from a treated solution of free-matrix sample analyte. To calculate the TYB, (namely, the analytical blank of the Youden plot), the analytical signal against increasing amounts of test portion (TP) should be plotted.

The value obtained from extrapolation to TP = 0 is the TYB, directly involved in the occurrence of constant bias in calibration (Fig. 4.5).

The TYB includes the signals due to the contributions of the sample except the analyte, that of the solvent and that due to the analyte-matrix interaction. SB instead contains only the signal due to the non-analyte in the solvent and the systematic error of negative correlation. Therefore, both the SB and the TBY also contain any constant errors due to the measurement system which is typical of the technique used and presumably not removable.

The SB and TBY values often differ as there is an interaction of the matrix with the analyte. This difference is called Youden white (Yb). If there is no analyte-matrix interaction YB should be practically zero and TBY and SB are equivalent or better not statistically different.

YB is a diagnostic parameter as it indicates sample-related errors.

The important relationship is:

$$\mathrm{TYB} - \mathrm{SB} = \mathrm{YB}$$

The overall functionality of MOSA is the relationship

$$\mathrm{Y} = \mathrm{m_M}\,\mathrm{W_r} + \mathrm{TBY}$$

And as we can see from Fig. 2 is an extension of the Youden curve and, that an unspiked sample signal response is equal to the MOSA intercept. Consequently, for a correct MOSA calculation the total Youden blank (TYB) must be subtracted.

Whenever possible, a multiple-point standardization is preferred, with results displayed as a calibration curve. A linear regression analysis can provide an equation for the standardization. A reagent blank corrects for any contribution to the signal from the reagents used in the analysis. The most common reagent blank is one in which an analyte-free sample is taken through the analysis. When a simple reagent blank does not compensate for all constant sources of determinate error, other types of blanks, such as the total Youden blank, can be used.

The standard addition line is an in situ normalization procedure for the proportional error. This normalization is due to the change of the slope of the MOSA curve with respect to the slope of the curve of the standard solutions. The two curves represent the responses due to the standards but the first in pure solvent and the second in the presence of the sample matrix. If the two lines are parallel and therefore have the same slope, there is not proportional error; otherwise, the ratio between the two slopes can be used to calculate the proportional error.

As long as we can ignore any constant bias due to interaction between the anlyte and the sample matrix, which is often the case, the accuracy of an analytical method will not suffer. IT is a good idea, however, to check for constant error sources of error before relying on either a calibration blank or a reagent blank.

A practical example with reference to Figure 4.5 will clarify the procedure.

The determination of Vitamin C in fruit juices by an amperometric method does not requires any sample pre-treatment except dilution. A series of standard were and by adding to 5 ml of sample, known amount of vitamin C standard and then diluting to 10 ml with water. The standard calibration curve (SC) was:

$$Y(mA) = 0.047C + 0.30$$

For the determination of the proportional error, a MOSA experiment was prepared preparing a series of samples to which know amount of standard were added except one that was simply diluted with water. Results are reported Fig. 4.5 (part B).

Regression line of MOSA

$$Y(mA) = 0.049C + 0.41$$

To calculate the TYB, different amounts of sample (2, 4. 6, 8 ml) were diluted to 10 ml with water and analyzed by the proposed method. The regression line reported in Fig. 2 (Part A) was obtained.

$$Y(mA) = 0.051C + 0.16$$

Where 0.16 is the TYB that is used to calculate the sample concentration:

$$C = \frac{Y - TYB}{m_s \times Sampleweight}$$

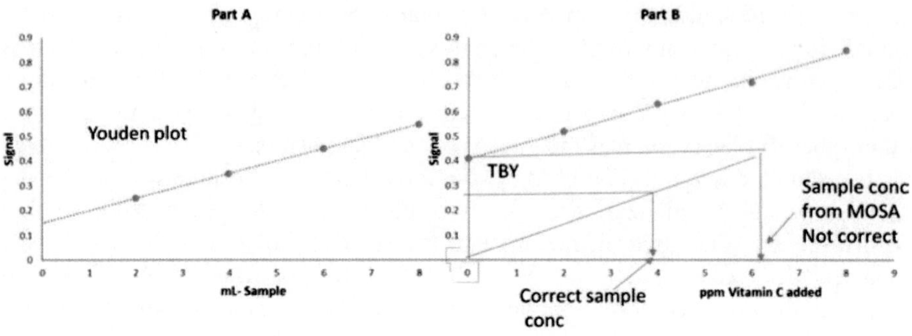

Fig. 4.5 Youden and MOSA plot

In part B of the Fig. 4.5 the correct concentration of Vitamin C in the sample (3.8 ppm) is reported, that is lower than the amount found considering only the MOSA results (6.2 ppm).

These results shows that, in this case, the constant bias due to the interaction between the analyte and the sample matrix cannot be ignored.

MOSA and EC were then carried out for the study of possible proportional bias in calibration.

The difference between these slopes is statistically non-significant and, therefore, the neglecting of proportional bias is acceptable. This ensures that conventional SC may be used for the quantization of Vitamin C in fruit juices.

For more information see:

1. Mario J. Cardone: J. Assoc. Off. Anal. Chem., 66,2, 1983, 1257–1282
2. Mario J. Cardone: J. Assoc. Off. Anal. Chem., 66,2, 1983, 1283–1294
3. Mario J. Cardone, Lehman: J. Assoc. Off. Anal. Chem., 68,2, 1985, 199–202
4. Mario J. Cardone, Susan A. Willavize, and Mark E. Lacy: Pharmaceutical Research, 7, 2 1990, 154–160

4.24 ANOVA- Analysis of Variance

The **analysis of variance (ANOVA) test** is used to compare the means of two or more independent samples and to test whether the differences between the means are statistically significant. It is a technique that can isolate and estimate variations associated with defined sources. ANOVA, which was developed by R. A. Fisher in the early 1920s, can be thought of as an extension of the *t* test for independent samples. However, a *t* test can compare only two means, whereas an ANOVA can comopare two or more means simultaneously.

In ANOVA, the *independent variable* is the categorical variable that defines the groups that are compared. The *dependent variable* is the measured variable whose means are being compared.

This is a procedure used to test for significant differences between means when there are *two or more means*. When two or more independent source of variation operate, the overall variance is the sum of the individual variances. For instance, in collaborative studies where different laboratories participate by analyzing the same sample, we face with two types of errors:

1. Between laboratories σ^2_b
2. Within laboratories σ^2_w

The total variance will be $\sigma^2 = \sigma^2_b + \sigma^2_w$

The test statistic in ANOVA is called the *F* statistic. The ***F* ratio** (or ***F* value**) is computed by dividing two variance estimates by each other. If the *F* ratio is statistically significant (i.e., if $p < .05$) and if there are three or more groups in the study, then a pairwise comparison is done to assess which two means are significantly different from each other.

The results of the ANOVA computations are often displayed in a summary table like the one displayed below.

4.24.1 The ANOVA Summary Table: General Format

Source	SS (Sum of squares)	df (degree of freedom)	MS (Mean squares)	F ratio (SS/df)	p
Between	SS_b	K – 1	MS_b	F-ratio	<0.05>
Within	SS_w	N-K	MS_w		
Total	SS_t	N – 1			

This table lists the sum of squares (SS), degrees of freedom (*df*), mean squares (MS), *F* ratio (*F*), and the level of significance (*p* level).

There are usually three type of sum of squares (SS):

SSt = total sum of squares that represent the variability of all the results around the total mean. It is the sum of the within groups sum of squares (SSw) and the between groups of sum od squares (SSb)

SSw = variability within the groups

SSb = between groups of sum of squares that is the average variability of the means of the groups around the total mean.

The Mean Squares (MS) are obtained by dividing the sum of square by the appropriate degree of freedom. Furthermore, by dividing the sum of squares by the appropriate degrees of freedom the F values are obtained. These will represent the proportion of variation between the groups compared to the variation within the groups. If the F value is higher than the F-critical value obtained by the appropriate table, the variation between the groups is statistically significant.

In practice, ANOVA deals with three types of variability that are called the **sum of squares (SS)**.

SS or Sum of Squares

- This is the measure of the variation around the mean. There are usually three different values of SS calculated:
  - SSG measures variation of the group means around the overall mean (Between Groups)
  - SSE measures the variation of each observation around its group mean (Within Groups)

Below examples OF the two most common type used in laboratory are reported: one-way ANOVA and two-way ANOVA with repeated measures or not. The numerical examples in the chapter guides you step by step through the analysis and interpretation of the data.

Example 1: Checking the Homogeneity of a Sample. (One-way Analysis of Variance – Two Replicate Only)
A certain amount of pork muscle is ground, mixed and then portioned into 100 g samples. The material must be certified for zinc content and we would like to know if the material is homogeneous. To this scope we formulate the null hypothesis (H_0) that the sample are homogeneous and the alternative hypothesis (H_1) that the samples are not. For this purpose, 10 randomly selected samples were analyzed in duplicate in the same laboratory in order to know if the material is homogeneous. To this scope we formulate the null hypothesis (H_0) that samples are homogeneous and the alternative hypothesis (H_1) that samples are not homogeneous

The results are displayed in the following table:

Samples	Results		Averages	SSj	$n(x_j - m)$
1	39.2	42.7	40.95	6.125	2.8322
2	41.2	43.3	42.25	2.205	12.4002
3	36.6	38.7	37.65	2.205	8.9042
4	39.5	39.0	39.25	0.125	0.5202
5	38.7	40.8	39.75	2.205	0.0002
6	42.8	39.0	40.90	7.220	2.5992
7	36.8	38.2	37.50	0.980	10.2151
8	39.3	40.0	39.65	0.245	0.0242
9	40.4	37.4	38.90	4.500	1.4792
10	39.6	42.0	40.80	2.880	2.1632
			M = 39.76	$SS_{w=}$ 28.69	$SS_b = 41.138$

SS_j = sum of squares of the difference among the single values of the analysis and the average of the of the two values. For instance, for the first sample we have that:

$$(39.2 - 40.95)^2 + (42.7 - 40.95)^2 = 6.125$$

By summing the different SS_j the sum of the squares within the groups (SS_w) is obtained.

The values of the last column $n(x_j - m)$ represent the square of the difference between the mean of the duplicates and the total average multiplied for the number (2 in this case) of analyses for each sample. For instance, for the first sample:

$$(40.95 - 39.76)^2 \times 2 = 2.8322$$

The sum of these value is SS among the group.

A one-way ANOVA tests the null hypothesis (*HO*) that states that all the groups represent populations that have the same means. The *alternative* hypothesis, HA (also called the *research hypothesis*), states that there is a statistically significant difference between at least two means.

In a summary table as reported in many statistical packages it will be like the one below:

	SS(Sum of squares)	df	MS	F ratio (SS/df)	Expected mean Sq.
Between samples	41.138	9	4.5709	1.59	$\sigma^2_0 + 2\sigma^2_1$.
Within samples	28.690	10	2.8690		σ^2_0
Total	69.828	19	3.6752		

Since the F ratio is 1.59 well below the critical value of F = 3.020 (from the F-table one tail 95% for 9 degree of freedom) we can retain the null hypothesis and say that there is no reason to say that our sample is non-homogeneous.

Retaining the null hypothesis means that the sample means are not significantly different from each other beyond what might be expected purely by chance, and we consider them as coming from the same population. v Example 2. **One-way analysis of variance (more replicates)**. Consider an experiment in which a reference material is analyzed each working day for a week, with four independent replicates each day. The following results are reported:

	Days				
Replicate	1	2	3	4	5
1	8.7	9.5	9.7	9.3	7.1
2	8.3	10.0	8.7	9.5	8.2
3	8.5	9.5	8.8	8.5	7.8
4	9.0	5.4	9.0	9.0	8.3
Sum	35.5	34.4	36.2	36.3	31.4

The null hypothesis is that there are no differences among the means of the five populations obtained in the 5 days. Excel or other software do the calculation and sort the following table.

Source	SS (Sum of squares)	df (degree of freedom)	MS (Mean squares)	F ratio (SS/df)	p
Between	2.26	5 – 1 = 4	0.565	9.26	<.0.05>
Within	0.92	y(4 – 1) = 15	0.92		
Total	SS_t	20 – 1 = 19	3.18		

If we choose the significance level α = 0.05, the F(4,15) value reported in the table is 3.06. Since the F value found is 9.26 we can conclude that the day averages differ by more that can be explained by the within –days experimental error.

Example 2: Comparison of Different Analysts (or Laboratories)
Below another example OF one-way ANOVA test concerning different analysts (or laboratories) analyse 6 times the same sample. In this case the lab responsible would like to know how precisely work the three analysts (or laboratories). He is interested in the type of error that occur between and within analysts (or laboratories). In this example we have an equal number of observations.

Three analyst in a laboratory analyze six times with the same method a meat sample for the content of protein. Results are presented in the following scheme:

	Analyst A	Analyst B	Analyst C			
1	20.83	29.17	21.66			
2	21.66	29.16	19.62			
3	19.16	26.62	19.61			
4	22.50	26.64	20.82			
5	20.84	20.83	23.30			
6	25.00	20.86	19.60			
Total	**156**	**184**	**148**			
Anova: Single factor						
SUMMARY						
Groups	*Count*	*Sum*	*Average*	*Variance*		
Analyst A	6	129.99	21.665	3.89447		
Analyst B	6	153.28	25.54667	14.54879		
Analyst C	6	124.61	20.76833	2.246497		
ANOVA						
Source of variation	*SS*	*df*	*MS*	*F*	*P-value*	*F crit*
Between groups	77.40763	2	38.70382	5.612027	0.015151	3.68232
Within groups	103.4488	15	6.896584			
Total	180.8564	17				

If we choose the significance level α = 0.05, the F(4,15) value reported in the table is 3.68 that is lower than the F found (5.61). The null hypothesis is rejected and that at least two sample means differ significantly from each other. When the null hypothesis is rejected, the next step is to conduct a **post hoc comparison**, in which all possible pairs of means are compared in order to find out which pair(s) of means differ(s).

4.25 Two-Way ANOVA

The above exercises are example of one-way ANOVA but there are several types of ANOVA tests. In this paragraph, a test called *two-way ANOVA is introduced.* It is used when there are two independent variable and the lab likes to test for an *interaction* between the two. Again, a numerical example is used to illustrate this statistical test. In general, when two or more independent variables are studied in ANOVA, the design is called a *factorial analysis of variance*.

A two-way ANOVA can be used to explore the effect of the two independent variables on the dependent variable. In addition, using the two-way ANOVA would allow us to study the interactions among all variables. In the factorial ANOVA test, the **interaction** refers to a situation where one or more levels of the independent variable have a different effect on the dependent variable when combined with another independent variable.

The *two-way* ANOVA test is designed to study the relationship between *two or more* independent variables and a dependent variable. One advantage of the two-way ANOVA is that it can reveal an *interaction* between the two independent variables. This interaction may not be apparent when a series of one-way ANOVA tests is conducted.

The total variation in a two-way ANOVA is partitioned into two main sources: the *within-groups* variation and the *between-groups* variation. The between group variation is further partitioned into three components: (a) the variation among the *row means*, (b) the variation among the *column means*, and (c) the variation due to the *interaction*.

Four mean squares (MS) are computed in a two-way ANOVA. Two are computed for the two *main effects* (the independent variables), one is computed for the *interaction*, and one for the *within*. Then, using the mean squares within (MSW) as the denominator, three *F* ratios are computed. These *F* ratios are found by dividing each of the three mean squares (MS*Row*, MS*Column*, and MS*Row X Column*) by MSW. As was the case with a one-way ANOVA, a summary table is used to display the two-way ANOVA summary information. The table includes the sum of squares, degrees of freedom, mean squares, *F* ratios, and *p* levels.

The general format of the Anova summary table is:

Source	SS (Sum of squares)	df (degree of freedom)	MS (Mean squares)	F ratio (SS/df)	p
Main effect					
Row	SS_R	Levels-1	MSr	Fr	
Column	SS_C	Levels-1	MSc	Fc	
Interaction	SS_{RC}	dfrow × dfc	MSrc	Frc	
Within groups	SS_w	N-K	MS_w		
Total	SS_t	N – 1			

A two-way ANOVA is conducted to test three null hypotheses about the effect of each of the two independent variables on the dependent variable and about the interaction between the two independent variables. The two independent variables (or *factors*) are referred to as the *row* variable and the *column* variable.

To test the three null hypotheses, three *F* ratios are calculated in a two-way ANOVA.

The three null hypotheses are:

*H*O(Row): There are no statistically significant differences among population means on the dependent measure for the first factor (the *row factor*).

*H*O(Column): There are no statistically significant differences among the population means on the dependent measure for the second factor (the *column factor*).

*H*O(Interaction): In the population, there is no statistically significant interaction between

Factor 1 and Factor 2 (the *row column interaction*).

4.25.1 The Two-Way ANOVA Summary Table

As was mentioned, the results of the computations of a two-way ANOVA test are presented in a summary table, similar to the table that is used for presenting the results of a one-way ANOVA. Each of the two factors in a two-way ANOVA, each of the two factors (which are also called the main effects) is associated with an F ratio. Similarly, the interaction is analyzed using its own F ratio. As in the one-way ANOVA, a summary table is used to display the results of the two-way ANOVA analyses (see Table 4.1).

Note, though, that there are three F ratios in the table, as opposed to a single F ratio in the one-way ANOVA. The three F ratios in a two-way ANOVA are designed to test the three null hypotheses. Table 4.1 also shows the computations of the degrees of freedom associated with each of the three F tests.

Again, an example will explain better the computations.

Three wine samples were analyzed in duplicates by 2 analysts using the same method. So, we are using 3 samples, 2 analysts and 2 analyses for a total of 60 measurements.

	Analyst 1		Analyst 2		Analyst 3	
	1	2	1	2	1	2
1	78.24	72.12	75.48	67.56	85.92	72.72
2	102.96	103.56	102.84	96.6	110.4	104.88
3	120.24	113.76	120.12	113.4	128.76	125.28
4	102	114.12	101.76	108.36	110.76	113.52
5	65.64	78.96	62.04	72	70.68	80.64
6	118.44	108.24	111.24	104.64	118.68	112.2
7	113.4	113.4	109.2	112.08	114.48	123.96
8	104.64	98.88	100.68	94.56	111.6	102.96
9	98.88	98.64	96.84	96.36	105.48	105.72
10	120.24	125.88	119.64	123.84	125.16	133.8

We formulate two null hypotheses, one for each factor (i.e., one for the samples, and one for the analysts). The first H0 is that the sample are homogeneous, while the alternative (H1) is that the samples are not. With the second H0, we hypothesize that the three analysts obtain similar results, while the alternative (H1) is that the analysts have different results.

A summary table can be obtained by many statistical packages. In MS Excel, this is quite straightforward. The result will be like the table below:

ANOVA						
Source of variation	*SS*	*df*	*MS*	*F*	*P-value*	*F crit*
Sample	16625.5	9	1847.28	70.38	1.06E-17	2.21
Analysts	723.580	2	361.790	13.78	5.68E-05	3.32
Interaction	51.2885	18	2.84936	0.11	0.9999	1.96
Within	787.414	30	26.2471			
Total	18187.8	59				

The interpretation of the results is similar to that of one-way ANOVA.

For instance, to check if the sample are homogeneous (the first null hypothesis), we compare the observed F value (70.38) with the critical one (Fcrit = 2.21) for 9 and 30 degrees of freedoms. Since the observed F values is bigger than the critical one, we reject the null hypothesis H0 and we conclude that the samples are significantly different at the level of 5%. Similarly, the F test for the Analyst has a F test of 13.78. This is again bigger than the critical F test. Again, we can reject the null hypothesis and conclude that also the analysts provides different results. Finally, the F-test for the interaction is equal to 0.11. This value is smaller than the critical F values. Accordingly, we cannot reject the null hypothesis. We conclude that based on the experimental data we could not observe a significant effect of the interaction between samples and analysts.

A further example is the following. Three analysts (A, B, and C) have to analyze two samples (I and II) in triplicates. The null hypotheses that we are interested to test are if the samples are similar, if the analysts get same results and if there is an interaction between samples and analysts. The results are summarized in the Table below:

	A	B	C
I	20.83	29.17	21.66
	21.66	29.16	19.62
	19.16	26.62	19.61
II	22.5	26.64	20.82
	20.84	20.83	23.3
	25	20.86	19.6

The resulting two ways ANOVA generates the following summary table:

ANOVA						
Source of variation	*SS*	*Df*	*MS*	*F*	*P-value*	*F crit*
Samples	2.8	1	2.8	0.7	0.42	4.75
Analysts	77.4	2	38.7	9.6	0.003	3.89
Interaction	52.0	2	26.0	6.4	0.013	3.89
Error	48.6	12	4.1			
Total	180.9	17				

The results show that the two samples are not statistically different ($F_{1,12} = 0.7 < F_{crit} = 4.75$), whereas there is a significant difference among the analysts ($F_{2,12} = 9.6 > F_{crit} = 3.89$). However, what is striking here is that there is a significant interaction between samples and analysts ($F_{2,12} = 6.4 > F_{crit} = 3.89$). When the interaction in a 2-ways anova table is significant, the researcher should focus only on this effect. In details, the interaction in this case means that each analyst has a different skill or attitude to analyze the two samples. This is highlighted in the interaction plot shown in the Figure below.

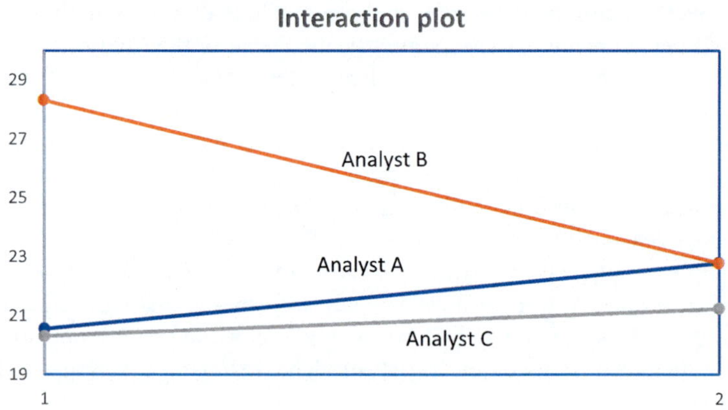

It is evident that the Analyst A is expressing a higher result for the sample II, whereas the Analyst B makes opposite, assigning the higher result to the sample I. Analyst C seems giving a similar result to both samples. Such differences among the analysts are greater in magnitude than the differences observed during replicates measurements. Accordingly, the ANOVA table indicates that the observed interaction is not a casual event, and that we can reject the null hypothesis (i.e., absence of interaction).

4.26 Meaning of p-Value

When observing different sample means it is not always possible to conclude that the population have different means. Maybe the population have the same mean and that the observed difference is only a simple coincidence. It is almost impossible to decide that the observed difference is a true difference or is just a coincidence of the random sampling. However, the question can be resolved recurring to probabilities.

If the population really has the same mean, which is the probability of observing a difference among sample means? The P value answer to this question. P value is a probability that can assume values from zero to one. More the P value is near to zero better we can conclude that the difference between sample means is unlikely to be a coincidence and the population have equal means. On the contrary, when P is more near to one the populations have different means. For instance, if the P value is 0.04, we can conclude that random sampling from a population would led to a difference smaller than that observed in 96% of the cases and larger than that observed in 4% of the cases.

It is important to note that many statistical packages report, by tradition, the value of $P < 0.05$ or $P > 0.05$ or they present a table like the one below

P values	
>0.05	Not significant
0.01 to 0.05	Significant
0.001 to 0.01	Very significant
<0.001	Extremely significant

Appendix

Shapiro-Wilk Test

Table 4.1 Coefficients

n=	2	3	4	5	6	7	8	9	10	11	12	13	14
a1	0.7071	0.7071	0.6872	0.6646	0.6431	0.6233	0.6052	0.5888	0.5739	0.5601	0.5475	0.5359	0.5251
a2			0.1677	0.2413	0.2806	0.3031	0.3164	0.3244	0.3291	0.3315	0.3325	0.3325	0.3318
a3					0.0875	0.1401	0.1743	0.1976	0.2141	0.2260	0.2347	0.2412	0.2460
a4							0.0561	0.0947	0.1224	0.1429	0.1586	0.1707	0.1802
a5									0.0399	0.0695	0.0922	0.1099	0.1240
a6											0.0803	0.0539	0.0727
a7													0.0240

Table 4.2 p-values

n\p	0.01	0.02	0.05	0.1	0.5	0.9	0.95	0.98	0.99
3	0.753	0.756	0.767	0.789	0.959	0.998	0.999	1.000	1.000
4	0.687	0.707	0.748	0.792	0.935	0.987	0.992	0.996	0.997
5	0.686	0.715	0.762	0.806	0.927	0.979	0.986	0.991	0.993
6	0.713	0.743	0.788	0.826	0.927	0.974	0.981	0.986	0.989
7	0.730	0.760	0.803	0.838	0.928	0.972	0.979	0.985	0.988
8	0.749	0.778	0.818	0.851	0.932	0.972	0.978	0.984	0.987
9	0.764	0.791	0.829	0.859	0.935	0.972	0.978	0.984	0.986
10	0.781	0.806	0.842	0.869	0.938	0.972	0.978	0.983	0.986
11	0.792	0.817	0.850	0.876	0.940	0.973	0.979	0.984	0.986
12	0.805	0.828	0.859	0.883	0.943	0.973	0.979	0.984	0.986
13	0.814	0.837	0.866	0.889	0.945	0.974	0.979	0.984	0.986
14	0.825	0.846	0.874	0.895	0.947	0.975	0.980	0.984	0.986
15	0.835	0.855	0.881	0.901	0.950	0.975	0.980	0.984	0.987
16	0.844	0.863	0.887	0.906	0.952	0.976	0.981	0.985	0.987
17	0.851	0.869	0.892	0.910	0.954	0.977	0.981	0.985	0.987
18	0.858	0.874	0.897	0.914	0.956	0.978	0.982	0.986	0.988
19	0.863	0.879	0.901	0.917	0.957	0.978	0.982	0.986	0.988
20	0.868	0.884	0.905	0.920	0.959	0.979	0.983	0.986	0.988
21	0.873	0.888	0.908	0.923	0.960	0.980	0.983	0.987	0.989
22	0.878	0.892	0.911	0.926	0.961	0.980	0.984	0.987	0.989
23	0.881	0.895	0.914	0.928	0.962	0.981	0.984	0.987	0.989
24	0.884	0.898	0.916	0.930	0.963	0.981	0.984	0.987	0.989
25	0.888	0.901	0.918	0.931	0.964	0.981	0.985	0.988	0.989
26	0.891	0.904	0.920	0.933	0.965	0.982	0.985	0.988	0.989
27	0.894	0.906	0.923	0.935	0.965	0.982	0.985	0.988	0.990
28	0.896	0.908	0.924	0.936	0.966	0.982	0.985	0.988	0.990
29	0.898	0.910	0.926	0.937	0.966	0.982	0.985	0.988	0.990
30	0.900	0.912	0.927	0.939	0.967	0.983	0.985	0.988	0.990
31	0.902	0.914	0.929	0.940	0.967	0.983	0.986	0.988	0.990
32	0.904	0.915	0.930	0.941	0.968	0.983	0.986	0.988	0.990
33	0.906	0.917	0.931	0.942	0.968	0.983	0.986	0.989	0.990
34	0.908	0.919	0.933	0.943	0.969	0.983	0.986	0.989	0.990
35	0.910	0.920	0.934	0.944	0.969	0.984	0.986	0.989	0.990
36	0.912	0.922	0.935	0.945	0.970	0.984	0.986	0.989	0.990
37	0.914	0.924	0.936	0.946	0.970	0.984	0.987	0.989	0.990
38	0.916	0.925	0.938	0.947	0.971	0.984	0.987	0.989	0.990
39	0.917	0.927	0.939	0.948	0.971	0.984	0.987	0.989	0.991
40	0.919	0.928	0.940	0.949	0.972	0.985	0.987	0.989	0.991
41	0.920	0.929	0.941	0.950	0.972	0.985	0.987	0.989	0.991
42	0.922	0.930	0.942	0.951	0.972	0.985	0.987	0.989	0.991
43	0.923	0.932	0.943	0.951	0.973	0.985	0.987	0.990	0.991
44	0.924	0.933	0.944	0.952	0.973	0.985	0.987	0.990	0.991
45	0.926	0.934	0.945	0.953	0.973	0.985	0.988	0.990	0.991
46	0.927	0.935	0.945	0.953	0.974	0.985	0.988	0.990	0.991
47	0.928	0.936	0.946	0.954	0.974	0.985	0.988	0.990	0.991
48	0.929	0.937	0.947	0.954	0.974	0.985	0.988	0.990	0.991
49	0.929	0.939	0.947	0.955	0.974	0.985	0.988	0.990	0.991
50	0.930	0.938	0.947	0.955	0.974	0.985	0.988	0.990	0.991

Grubb Statistic Values

Values of Grubbs Statistic (G)

Number of Observations n	Confidence Level (%)					
	99.9	99.5	99	97.5	95	90
3	1.155	1.155	1.155	1.155	1.153	1.148
4	1.499	1.496	1.492	1.481	1.463	1.425
5	1.780	1.764	1.749	1.715	1.672	1.602
6	2.011	1.973	1.944	1.887	1.822	1.729
7	2.201	2.139	2.097	2.020	1.938	1.828
8	2.358	2.274	2.221	2.126	2.032	1.909
9	2.492	2.387	2.323	2.215	2.110	1.977
10	2.606	2.482	2.410	2.290	2.176	2.036
11	2.705	2.564	2.485	2.355	2.234	2.088
12	2.791	2.636	2.550	2.412	2.285	2.134
13	2.867	2.699	2.607	2.462	2.331	2.175
14	2.935	2.755	2.659	2.507	2.371	2.213
15	2.997	2.806	2.705	2.549	2.409	2.247
16	3.052	2.852	2.747	2.585	2.443	2.279
17	3.103	2.894	2.785	2.620	2.475	2.309
18	3.149	2.932	2.821	2.651	2.504	2.335
19	3.191	2.968	2.854	2.681	2.532	2.361
20	3.230	3.001	2.884	2.709	2.557	2.385
30	3.507	3.236	3.103	2.908	2.745	2.563
40	3.673	3.381	3.240	3.036	2.866	2.682
50	3.789	3.483	3.336	3.128	2.956	2.768
60	3.874	3.560	3.411	3.199	3.025	2.837
70	3.942	3.622	3.471	3.257	3.082	2.893
80	3.998	3.673	3.521	3.305	3.130	2.940
90	4.044	3.716	3.563	3.347	3.171	2.981
100	4.084	3.754	3.600	3.383	3.207	3.017

Source: ASTM E178-00, "Standard Practice for Dealing with Outlying Observations"

F-table for α = 0.05

	F-table of Critical Values of α = 0.05 for F(df1, df2)																		
	DF1=1	2	3	4	5	6	7	8	9	10	12	15	20	24	30	40	60	120	∞
DF2=1	161.45	199.50	215.71	224.58	230.16	233.99	236.77	238.88	240.54	241.88	243.91	245.95	248.01	249.05	250.10	251.14	252.20	253.25	254.31
2	18.51	19.00	19.16	19.25	19.30	19.33	19.35	19.37	19.38	19.40	19.41	19.43	19.45	19.45	19.46	19.47	19.48	19.49	19.50
3	10.13	9.55	9.28	9.12	9.01	8.94	8.89	8.85	8.81	8.79	8.74	8.70	8.66	8.64	8.62	8.59	8.57	8.55	8.53
4	7.71	6.94	6.59	6.39	6.26	6.16	6.09	6.04	6.00	5.96	5.91	5.86	5.80	5.77	5.75	5.72	5.69	5.66	5.63
5	6.61	5.79	5.41	5.19	5.05	4.95	4.88	4.82	4.77	4.74	4.68	4.62	4.56	4.53	4.50	4.46	4.43	4.40	4.37
6	5.99	5.14	4.76	4.53	4.39	4.28	4.21	4.15	4.10	4.06	4.00	3.94	3.87	3.84	3.81	3.77	3.74	3.70	3.67
7	5.59	4.74	4.35	4.12	3.97	3.87	3.79	3.73	3.68	3.64	3.57	3.51	3.44	3.41	3.38	3.34	3.30	3.27	3.23
8	5.32	4.46	4.07	3.84	3.69	3.58	3.50	3.44	3.39	3.35	3.28	3.22	3.15	3.12	3.08	3.04	3.01	2.97	2.93
9	5.12	4.26	3.86	3.63	3.48	3.37	3.29	3.23	3.18	3.14	3.07	3.01	2.94	2.90	2.86	2.83	2.79	2.75	2.71
10	4.96	4.10	3.71	3.48	3.33	3.22	3.14	3.07	3.02	2.98	2.91	2.85	2.77	2.74	2.70	2.66	2.62	2.58	2.54
11	4.84	3.98	3.59	3.36	3.20	3.09	3.01	2.95	2.90	2.85	2.79	2.72	2.65	2.61	2.57	2.53	2.49	2.45	2.40
12	4.75	3.89	3.49	3.26	3.11	3.00	2.91	2.85	2.80	2.75	2.69	2.62	2.54	2.51	2.47	2.43	2.38	2.34	2.30
13	4.67	3.81	3.41	3.18	3.03	2.92	2.83	2.77	2.71	2.67	2.60	2.53	2.46	2.42	2.38	2.34	2.30	2.25	2.21
14	4.60	3.74	3.34	3.11	2.96	2.85	2.76	2.70	2.65	2.60	2.53	2.46	2.39	2.35	2.31	2.27	2.22	2.18	2.13
15	4.54	3.68	3.29	3.06	2.90	2.79	2.71	2.64	2.59	2.54	2.48	2.40	2.33	2.29	2.25	2.20	2.16	2.11	2.07
16	4.49	3.63	3.24	3.01	2.85	2.74	2.66	2.59	2.54	2.49	2.42	2.35	2.28	2.24	2.19	2.15	2.11	2.06	2.01
17	4.45	3.59	3.20	2.96	2.81	2.70	2.61	2.55	2.49	2.45	2.38	2.31	2.23	2.19	2.15	2.10	2.06	2.01	1.96
18	4.41	3.55	3.16	2.93	2.77	2.66	2.58	2.51	2.46	2.41	2.34	2.27	2.19	2.15	2.11	2.06	2.02	1.97	1.92
19	4.38	3.52	3.13	2.90	2.74	2.63	2.54	2.48	2.42	2.38	2.31	2.23	2.16	2.11	2.07	2.03	1.98	1.93	1.88
20	4.35	3.49	3.10	2.87	2.71	2.60	2.51	2.45	2.39	2.35	2.28	2.20	2.12	2.08	2.04	1.99	1.95	1.90	1.84
21	4.32	3.47	3.07	2.84	2.68	2.57	2.49	2.42	2.37	2.32	2.25	2.18	2.10	2.05	2.01	1.96	1.92	1.87	1.81
22	4.30	3.44	3.05	2.82	2.66	2.55	2.46	2.40	2.34	2.30	2.23	2.15	2.07	2.03	1.98	1.94	1.89	1.84	1.78
23	4.28	3.42	3.03	2.80	2.64	2.53	2.44	2.37	2.32	2.27	2.20	2.13	2.05	2.01	1.96	1.91	1.86	1.81	1.76
24	4.26	3.40	3.01	2.78	2.62	2.51	2.42	2.36	2.30	2.25	2.18	2.11	2.03	1.98	1.94	1.89	1.84	1.79	1.73
25	4.24	3.39	2.99	2.76	2.60	2.49	2.40	2.34	2.28	2.24	2.16	2.09	2.01	1.96	1.92	1.87	1.82	1.77	1.71
26	4.23	3.37	2.98	2.74	2.59	2.47	2.39	2.32	2.27	2.22	2.15	2.07	1.99	1.95	1.90	1.85	1.80	1.75	1.69
27	4.21	3.35	2.96	2.73	2.57	2.46	2.37	2.31	2.25	2.20	2.13	2.06	1.97	1.93	1.88	1.84	1.79	1.73	1.67
28	4.20	3.34	2.95	2.71	2.56	2.45	2.36	2.29	2.24	2.19	2.12	2.04	1.96	1.91	1.87	1.82	1.77	1.71	1.65
29	4.18	3.33	2.93	2.70	2.55	2.43	2.35	2.28	2.22	2.18	2.10	2.03	1.94	1.90	1.85	1.81	1.75	1.70	1.64
30	4.17	3.32	2.92	2.69	2.53	2.42	2.33	2.27	2.21	2.16	2.09	2.01	1.93	1.89	1.84	1.79	1.74	1.68	1.62
40	4.08	3.23	2.84	2.61	2.45	2.34	2.25	2.18	2.12	2.08	2.00	1.92	1.84	1.79	1.74	1.69	1.64	1.58	1.51
60	4.00	3.15	2.76	2.53	2.37	2.25	2.17	2.10	2.04	1.99	1.92	1.84	1.75	1.70	1.65	1.59	1.53	1.47	1.39
120	3.92	3.07	2.68	2.45	2.29	2.18	2.09	2.02	1.96	1.91	1.83	1.75	1.66	1.61	1.55	1.50	1.43	1.35	1.25
∞	3.84	3.00	2.60	2.37	2.21	2.10	2.01	1.94	1.88	1.83	1.75	1.67	1.57	1.52	1.46	1.39	1.32	1.22	1.00

F-table for α = 0.10

F-table of Critical Values of α = 0.10 for F(df1, df2)																			
	DF1=1	2	3	4	5	6	7	8	9	10	12	15	20	24	30	40	60	120	∞
DF2=1	39.86	49.50	53.59	55.83	57.24	58.20	58.91	59.44	59.86	60.19	60.71	61.22	61.74	62.00	62.26	62.53	62.79	63.06	63.33
2	8.53	9.00	9.16	9.24	9.29	9.33	9.35	9.37	9.38	9.39	9.41	9.42	9.44	9.45	9.46	9.47	9.47	9.48	9.49
3	5.54	5.46	5.39	5.34	5.31	5.28	5.27	5.25	5.24	5.23	5.22	5.20	5.18	5.18	5.17	5.16	5.15	5.14	5.13
4	4.54	4.32	4.19	4.11	4.05	4.01	3.98	3.95	3.94	3.92	3.90	3.87	3.84	3.83	3.82	3.80	3.79	3.78	3.76
5	4.06	3.78	3.62	3.52	3.45	3.40	3.37	3.34	3.32	3.30	3.27	3.24	3.21	3.19	3.17	3.16	3.14	3.12	3.11
6	3.78	3.46	3.29	3.18	3.11	3.05	3.01	2.98	2.96	2.94	2.90	2.87	2.84	2.82	2.80	2.78	2.76	2.74	2.72
7	3.59	3.26	3.07	2.96	2.88	2.83	2.78	2.75	2.72	2.70	2.67	2.63	2.59	2.58	2.56	2.54	2.51	2.49	2.47
8	3.46	3.11	2.92	2.81	2.73	2.67	2.62	2.59	2.56	2.54	2.50	2.46	2.42	2.40	2.38	2.36	2.34	2.32	2.29
9	3.36	3.01	2.81	2.69	2.61	2.55	2.51	2.47	2.44	2.42	2.38	2.34	2.30	2.28	2.25	2.23	2.21	2.18	2.16
10	3.29	2.92	2.73	2.61	2.52	2.46	2.41	2.38	2.35	2.32	2.28	2.24	2.20	2.18	2.16	2.13	2.11	2.08	2.06
11	3.23	2.86	2.66	2.54	2.45	2.39	2.34	2.30	2.27	2.25	2.21	2.17	2.12	2.10	2.08	2.05	2.03	2.00	1.97
12	3.18	2.81	2.61	2.48	2.39	2.33	2.28	2.24	2.21	2.19	2.15	2.10	2.06	2.04	2.01	1.99	1.96	1.93	1.90
13	3.14	2.76	2.56	2.43	2.35	2.28	2.23	2.20	2.16	2.14	2.10	2.05	2.01	1.98	1.96	1.93	1.90	1.88	1.85
14	3.10	2.73	2.52	2.39	2.31	2.24	2.19	2.15	2.12	2.10	2.05	2.01	1.96	1.94	1.91	1.89	1.86	1.83	1.80
15	3.07	2.70	2.49	2.36	2.27	2.21	2.16	2.12	2.09	2.06	2.02	1.97	1.92	1.90	1.87	1.85	1.82	1.79	1.76
16	3.05	2.67	2.46	2.33	2.24	2.18	2.13	2.09	2.06	2.03	1.99	1.94	1.89	1.87	1.84	1.81	1.78	1.75	1.72
17	3.03	2.64	2.44	2.31	2.22	2.15	2.10	2.06	2.03	2.00	1.96	1.91	1.86	1.84	1.81	1.78	1.75	1.72	1.69
18	3.01	2.62	2.42	2.29	2.20	2.13	2.08	2.04	2.00	1.98	1.93	1.89	1.84	1.81	1.78	1.75	1.72	1.69	1.66
19	2.99	2.61	2.40	2.27	2.18	2.11	2.06	2.02	1.98	1.96	1.91	1.86	1.81	1.79	1.76	1.73	1.70	1.67	1.63
20	2.97	2.59	2.38	2.25	2.16	2.09	2.04	2.00	1.96	1.94	1.89	1.84	1.79	1.77	1.74	1.71	1.68	1.64	1.61
21	2.96	2.57	2.36	2.23	2.14	2.08	2.02	1.98	1.95	1.92	1.87	1.83	1.78	1.75	1.72	1.69	1.66	1.62	1.59
22	2.95	2.56	2.35	2.22	2.13	2.06	2.01	1.97	1.93	1.90	1.86	1.81	1.76	1.73	1.70	1.67	1.64	1.60	1.57
23	2.94	2.55	2.34	2.21	2.11	2.05	1.99	1.95	1.92	1.89	1.84	1.80	1.74	1.72	1.69	1.66	1.62	1.59	1.55
24	2.93	2.54	2.33	2.19	2.10	2.04	1.98	1.94	1.91	1.88	1.83	1.78	1.73	1.70	1.67	1.64	1.61	1.57	1.53
25	2.92	2.53	2.32	2.18	2.09	2.02	1.97	1.93	1.89	1.87	1.82	1.77	1.72	1.69	1.66	1.63	1.59	1.56	1.52
26	2.91	2.52	2.31	2.17	2.08	2.01	1.96	1.92	1.88	1.86	1.81	1.76	1.71	1.68	1.65	1.61	1.58	1.54	1.50
27	2.90	2.51	2.30	2.17	2.07	2.00	1.95	1.91	1.87	1.85	1.80	1.75	1.70	1.67	1.64	1.60	1.57	1.53	1.49
28	2.89	2.50	2.29	2.16	2.06	2.00	1.94	1.90	1.87	1.84	1.79	1.74	1.69	1.66	1.63	1.59	1.56	1.52	1.48
29	2.89	2.50	2.28	2.15	2.06	1.99	1.93	1.89	1.86	1.83	1.78	1.73	1.68	1.65	1.62	1.58	1.55	1.51	1.47
30	2.88	2.49	2.28	2.14	2.05	1.98	1.93	1.88	1.85	1.82	1.77	1.72	1.67	1.64	1.61	1.57	1.54	1.50	1.46
40	2.84	2.44	2.23	2.09	2.00	1.93	1.87	1.83	1.79	1.76	1.71	1.66	1.61	1.57	1.54	1.51	1.47	1.42	1.38
60	2.79	2.39	2.18	2.04	1.95	1.87	1.82	1.77	1.74	1.71	1.66	1.60	1.54	1.51	1.48	1.44	1.40	1.35	1.29
120	2.75	2.35	2.13	1.99	1.90	1.82	1.77	1.72	1.68	1.65	1.60	1.55	1.48	1.45	1.41	1.37	1.32	1.26	1.19
∞	2.71	2.30	2.08	1.94	1.85	1.77	1.72	1.67	1.63	1.60	1.55	1.49	1.42	1.38	1.34	1.30	1.24	1.17	1.00

Student t-table

	P						
one-tail	0.1	0.05	0.025	0.01	0.005	0.001	0.0005
two-tails	0.2	0.1	0.05	0.02	0.01	0.002	0.001
DF							
1	3.078	6.314	12.706	31.821	63.656	318.289	636.578
2	1.886	2.92	4.303	6.965	9.925	22.328	31.6
3	1.638	2.353	3.182	4.541	5.841	10.214	12.924
4	1.533	2.132	2.776	3.747	4.604	7.173	8.61
5	1.476	2.015	2.571	3.365	4.032	5.894	6.869
6	1.44	1.943	2.447	3.143	3.707	5.208	5.959
7	1.415	1.895	2.365	2.998	3.499	4.785	5.408
8	1.397	1.86	2.306	2.896	3.355	4.501	5.041
9	1.383	1.833	2.262	2.821	3.25	4.297	4.781
10	1.372	1.812	2.228	2.764	3.169	4.144	4.587
11	1.363	1.796	2.201	2.718	3.106	4.025	4.437
12	1.356	1.782	2.179	2.681	3.055	3.93	4.318
13	1.35	1.771	2.16	2.65	3.012	3.852	4.221
14	1.345	1.761	2.145	2.624	2.977	3.787	4.14
15	1.341	1.753	2.131	2.602	2.947	3.733	4.073
16	1.337	1.746	2.12	2.583	2.921	3.686	4.015
17	1.333	1.74	2.11	2.567	2.898	3.646	3.965
18	1.33	1.734	2.101	2.552	2.878	3.61	3.922
19	1.328	1.729	2.093	2.539	2.861	3.579	3.883
20	1.325	1.725	2.086	2.528	2.845	3.552	3.85
21	1.323	1.721	2.08	2.518	2.831	3.527	3.819
22	1.321	1.717	2.074	2.508	2.819	3.505	3.792
23	1.319	1.714	2.069	2.5	2.807	3.485	3.768
24	1.318	1.711	2.064	2.492	2.797	3.467	3.745
25	1.316	1.708	2.06	2.485	2.787	3.45	3.725
26	1.315	1.706	2.056	2.479	2.779	3.435	3.707
27	1.314	1.703	2.052	2.473	2.771	3.421	3.689
28	1.313	1.701	2.048	2.467	2.763	3.408	3.674
29	1.311	1.699	2.045	2.462	2.756	3.396	3.66
30	1.31	1.697	2.042	2.457	2.75	3.385	3.646
60	1.296	1.671	2	2.39	2.66	3.232	3.46
120	1.289	1.658	1.98	2.358	2.617	3.16	3.373
1000	1.282	1.646	1.962	2.33	2.581	3.098	3.3
Inf	1.282	1.645	1.96	2.326	2.576	3.091	3.291

Chapter 5
Uncertainty Measurements

It is well known that many analytical methods, internationally recognized, report the form of presentation of the result and specify limits for the results. In these situation laboratories are recommended to estimate uncertainty since experience shows that limits imposed by standard official methods are not sufficient to ensure results that are fit for purpose.

The revised International Vocabulary of Metrology (VIM) and General Terms in Metrology defines uncertainty as the "non-negative parameter characterizing the dispersion of the quantity values being attributed to the measurand, based on the information used". The measurand is defined as "the particular quantity subject to measurement". According to this definition, it is clear that the estimation of uncertainty is essential in the interpretation of the results and, in particular, their comparison with regulatory limits of conformity. Hence, ISO 17025 requires, since 2001, for testing laboratories to apply procedures for estimating uncertainty and to be able, if necessary, to be associated with the results returned. In fact, the point 7.2.1.1 of the ISO 17025: 2017 reports that "The laboratory shall use appropriate methods and procedures for all laboratory activities and, where appropriate, for evaluation of the measurement uncertainty as well as statistical techniques for analysis of data."

So, uncertainty is the parameter associated with the results of a measurement may be a standard deviation called standard measurement uncertainty having a stated coverage probability. In other words, the uncertainty provides an interval within which the value of the measurand can be found with a stated probability. Usually the small letter *u* denotes uncertainty but it is important to recognize the different terms of uncertainty:

1. Standard measurement uncertainty – $u\ (x_i)$
2. Combined standard measurement uncertainty – $u_c\ (y)$
3. Expanded measurement uncertainty – $U = k\ u_c$

S. Mannino, *Guidelines for Laboratory Quality Managers*, Integrating Food Science and Engineering Knowledge Into the Food Chain 14,
https://doi.org/10.1007/978-3-031-11724-4_5

The first one is usually expressed in standard deviation for the quantity x_i, the second one is a mathematical combination of several individual standard measurement uncertainties, and, finally the third one is simply obtained by multiplying the combined standard deviation by *a coverage factor (k)*, usually 2.

Having in mind these definitions, it is important to know and/or identify all sources of measurement uncertainty with the final goal to reduce the total uncertainty and to be in acceptable limits required by the customer. It is worthwhile to remember that many international and national legislations require the use of analytical procedure that can provide results with a fixed uncertainty of measurement.

A result, uncertainty is usually expressed as a number followed by a ± sign. Example: 10 ± 1 mg/L at a level of confidence of 95% tell us that the measurement is between 9 and 11 mg/L. Attention should be paid not to confuse uncertainty with error, where the latter is difference between the measured value and the "true value".

Uncertainty is important since tell us how good is the result and it is used for instance in calibration or in verifying tolerances.

In the laboratory, operation errors and uncertainties come from many sources but the mains are:

- The measuring instrument
- The sample
- The measuring process (method phases)
- External uncertainties like the instrument calibration
- Technician skill
- Sampling
- Environmental conditions

Now we will see how to calculate the "uncertainty of measurement" (MU).

5.1 Approaches to Estimate Measurement Uncertainty

There are a number of different approaches to estimate measurement uncertainty (MU) but the most popular are the bottom-up or top-down calculations.

The bottom-up approach described in the ISO/IEC Guide of the Expression of Uncertainty in Measurement", commonly called GUM, combines all the uncertainties associated with all individual operations of an analytical procedure and then calculate the combined uncertainty that is multiplied by a coverage factor (95% confidence range) to obtain the expanded uncertainty.

In 1995, EURACHEM and CITAC (Co-operation on International Traceability in Analytical Chemistry) published the QUAM-guide (Quantifying Uncertainty in Analytical Measurement), which was revised in 2000 and 2012. It provides guidance for both approach and numerous worked examples.

The GUM was created to harmonize the procedures used for measurement uncertainty evaluation. The QUAM-guide was written to help analytical chemists to use the approach described in GUM, including examples of analytical chemistry.

5.1.1 The Bottom Up Approach

In the bottom-up approach, to calculate the uncertainty of a measurement, all the sources of uncertainty in the measurement must be identified. After estimation of the size of the uncertainty from each source, the individual uncertainties should be combined to give an overall result.

The relevant uncertainty sources are shown in the cause and effect diagram depicted below (Fig. 5.1):

Cause and effect diagrams are used to brainstorm potential factors to be investigated in a design experiment. These chart are sometimes referred as fishbone diagram because of their form: causes are listed on lines with branches similar to branches of a fish spine. They help to identify all the input variable that could influence the measurement.

However, the causes listed are potential causes that may not really contribute to the overall uncertainty. Data will lead to the final judgement.

In most measurement situations, uncertainty evaluations called of types A and B are needed.

Type A evaluations – uncertainty estimates using statistics (usually from repeated readings).

Type B evaluations – uncertainty estimates from any other information. This could be information from past experience of the measurements, from calibration certificates, manufacturer's specifications, from calculations, from published information, and from common sense.

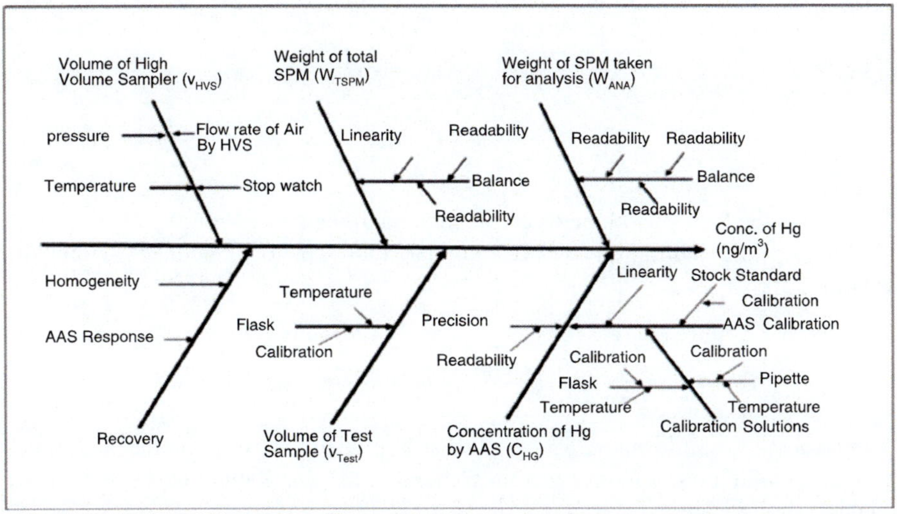

Fig. 5.1 Cause and effect diagram ('fishbone diagram') visualizing examples for possible MU contributions

However, the applied probability function (e.g. normal distribution) the degree of freedom and the coverage factors should also be included when needed.

5.1.1.1 How to Calculate the Standard Uncertainty for a Type A Evaluation

Type A evaluation is usually a standard deviation od the mean of repeated measurements performed under repeatability conditions. This has already been described. However, when a set of result have been obtained from several measurements we usually obtain the mean and we estimate the standard deviation with the usual formula. The estimated standard uncertainty (u) of the mean is calculate from:

$$u = \frac{s}{\sqrt{n}}$$

Where n is the number of measurements. U is often called the standard error of the mean.

5.1.1.2 How to Calculate the Standard Uncertainty for a Type B Evaluation

If we use other means than statistical treatment of repeated measurement results for calculating uncertainties, the so called type B estimated should be used. This is the case when we use certificate of reference materials, specification of instruments, volumetric flasks or pipette and son on.

In this type of evaluation, the type of distribution should be chosen. The rectangular distribution is the more common and the standard uncertainty can be calculating from:

$$u = \frac{a}{\sqrt{3}}$$

Where a is the half – width between the upper and the lower limit.

Another distribution is the rectangular that can be calculated from the relationship:

$$u = \frac{a}{\sqrt{6}}$$

In many cases, information about the distribution are missing and the analyst has to choose which one should be used. Generally, the distribution that overestimate the uncertainty should be used.

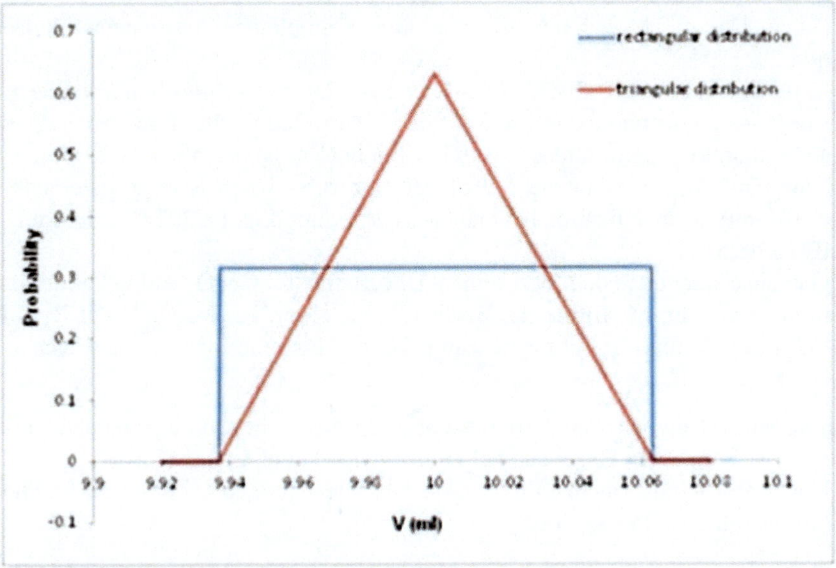

Fig. 5.2 Triangular and rectangular distribution

At this point, the uncertainties should be converted in the same units and then combined.

Individual standard uncertainties calculated by Type A or Type B evaluations can be combined validly by 'summation in quadrature' (also known as 'root sum of the squares'). The result of this is called the *combined standard uncertainty*:

$$Combined\,uncertainty = u = \sqrt{a^2 + b^2 + c^2 + \ldots etc}$$

Multiplying the standard uncertainty by a coverage factor we obtain the expanded uncertainty that assumes the symbol U; so

$$U = k\,u$$

Where K assume usually the value of 2 indicating a level of confidence of approximately of 95%.

At this point is possible to express the analytical answer including the uncertainty figure, e.g. 10 ± 1 mg/L followed by the following statement "the reported uncertainty is based on standard uncertainty multiplied by a coverage factor of 2, providing a level of confidence of approximately 95%."

In the last few decades, numerous guides have been published on the calculation of measurement uncertainty. For more examples refer to the following publications:

- *ILAC G8:09/2019 Guidelines on Decision Rules and Statements of* CITAC Guide: Use of Uncertainty Information in Compliance Assessment (2nd edition 2021)
- *ILAC G8:09/2019 G8:09/2019 Guidelines on Decision Rules and Statements of Conformity* This publication was extensively revised by the ILAC Accreditation and Laboratory Committees to provide guidance to laboratories, assessors, regulators and customers in the use of decision rules when issuing statements of conformity to specifications or standards as required in the 2017 edition of ISO/IEC 17025.
- Guidance Document on Measurement Uncertainty for GMO Testing Laboratories by S. Trapman, M. Burns, H. Broll, R. Macarthur, R. Wood, J. Zel By GRC European Commission Joint Research Centre- Institute for Reference Materials and Measurements

Furthermore, many softwares are present in the web that can help to calculate the uncertainty budget but Excel can help in the calculation.

However, we must say that this approach has been judged not practical for chemical measurements. In fact, it is:

- Difficult to write an equation that completely describe the measurement system including all influence factors i.e.– sample clean-up conditions, recovery of analyte from matrix, instrument conditions, interferences… and so on
- Evaluate the uncertainties associated with all parameters in the equation – Type A: statistical evaluation, Type B: any other data (certificates, instrument specifications, etc)
- Difficulties in evaluating individual uncertainty components
- Express all uncertainties as standard deviations
- Combine using mathematical rules for the combination of variances

In practice, the process is too time consuming and not easy to run in routine testing laboratories, so, the top-down approach is recommended to chemical laboratories.

5.1.2 Top Down Approach

According to the NORDTEST approach "Report TR 537, Handbook for calculation of measurement uncertainty in environmental laboratories (available from www.nordtest.info)" the combined measurement uncertainty can be broken down into two main components:

1. the within-laboratory reproducibility (intermediate precision) sRw and
2. the uncertainty due to possible laboratory bias u(bias).

Both of these can be conveniently estimated from validation and quality control data, thus significantly reducing the need for performing dedicated experiments for estimating detailed uncertainty contributions and thereby making uncertainty estimation easier for routine laboratories. An additional merit of this uncertainty estimation approach is that it reduces the danger of underestimating the uncertainty, which continues to be a problem at routine laboratories.

Its ease of use and the possibility to utilize available data from validation, and quality control have made the top-down approach popular among routine laboratories.

Here, a (two) worked example (s) on how to use the top-down approach to estimate MU is reported. In particular, we will use "the Nordtest approach" based on validation and quality control data, usually present in the laboratory.

Since then, several 'top-down' approaches have been presented for measurement uncertainty evaluation. Some of them are based on inter-laboratory data while others are based on intra-laboratory data.

5.1.2.1 Top-Down' Requirements

- The best available estimate of precision: – from validation studies or ongoing QC
- The best available estimate of bias and its uncertainty including method and laboratory bias
- In-house/inter-laboratory studies – ongoing internal quality control (IQC) data – proficiency testing data
- Look at the variation in method outputs (i.e. results) rather than method inputs
- Cover method scope – matrix, analyte concentration

5.2 Case Study – Determination of Cholesterol in Animal and Vegetable Fats and Oils

The method involves an extraction/clean-up step followed by quantification by GC-FID by using an internal standard for calibration. Analyses on reference samples were performed by two different analyst by analyzing in triplicate independent samples over five different days. In each case the repeatability standard deviation (s_r) and the intermediate precision (s_i) was calculated by using the analysis of variance (ANOVA). The results are reported in the table below.

Results from precision study

		Repeatability		Intermediate precision	
Sample type	Mean (x)	s_r	$s_{r/x}$	s_i	$S_{i/x}$
Anhydrous milk fat CRM	269.3	1.69	0.00628	2.93	0.0019
Spiked olive oil	106.2	0.840	0.00791	1.44	0.0135
Spiked corn oil	70.30	0.420	0.00597	0.73	0.0104
Pork and beef fat CRM	128.1	0.935	0.00730	1.62	0.0126
Pooled values		1.07	0.00691	1.86	0.0019

Precision data were calculated as usual. However, in this example the intermediate precision is used since it covers more sources of uncertainty than repeatability. In this case the pooled value of 0.0019 is used since samples are similar and the relative standard deviation of the four sample seems not very different.

At this point, bias can be obtained by validation or proficiency testing data if available. Validation data are usually obtained by using certified reference material or spiked samples. Our goal is to verify if bias is significant compared to the method precision and if his uncertainty is part of the overall uncertainty evaluation. Recovery should also be considered.

If bias is included in uncertainty estimates the following relation can be used:

$$\frac{u(R_m)}{R_m} = \sqrt{\left(\frac{u(C_{oss})}{C_{oss}}\right)^2 + \left(\frac{u(C_{cert})}{C_{cert}}\right)^2}$$

Where $u(C_{cert})$ is the value reported in the CRM certificate and $u(C_{oss})$ the value obtained from the analysis of the CRM and $u(R_m)$ recovery uncertainty.

If the bias is not significant, recovery is not significantly different from 100% – assume Rm = 1 with an uncertainty, u(Rm) – uncertainty on bias estimate included, even if bias itself is not significant.

If the bias is significant, some methodology to reduce bias should be found. u(Rm) should be included in uncertainty estimate for corrected results.

In our case we use recovery data obtained by analyzing 3 times a certified reference material and 7 different but similar matrices spiked with different level of the analyte. Data are reported in the following table

Results are available from the analysis of a reference material (anhydrous milk fat reference material CRM 164)

Mean (mg/100 g) ( )	269.33
Standard deviation (mg/100 g)	1.692
Number of replicates	11
Certified cholesterol content (mg/100 g) ( )	274.7 ± 9* *Uncertainty in certified value given at the 95% confidence level

(Rm) and uncertainty calculation from recovery data and estimation of the uncertainty u(Rm)

$$R_m = \frac{C_{oss}}{C_{cert}} = \frac{269.33}{274.7} = 0.98$$

U(Rm) has contribution from:

- the uncertainty of the certified value of CRM
- the uncertainty in the mean of the laboratory results u(C).

$$u(C_{cert}) = \frac{9.0}{1.96} = 4.59$$

While

$$u(C_{oss}) = \frac{1.692}{\sqrt{11}} = 0.51$$

So u(Rm)

$$\frac{\mathbf{u(R_m)}}{\mathbf{0.98}} = \sqrt{\left(\frac{\mathbf{0.51}}{\mathbf{269.33}}\right)^2 + \left(\frac{\mathbf{4.59}}{\mathbf{247.7}}\right)^2}$$

So $u(R_m)$ =0.016. This means that the recovery is estimated as 0.98 with a standard uncertainty of 0.016.

To see if the method recovery presents a significant bias we use the ratio.

$\frac{1-R_m}{u(R_m)}$ that we compare with K where K is equal to 2 for a confidence interval of 95%.

$$\frac{1-0.98}{0.016} = 1.19$$

Since 1.19 < 2 there is no evidence that the recovery is significantly different from 1 and no reason to correct experimental results for incomplete recovery.

Matrix effects u(Rs): Relative standard deviation

Summary of recovery data for cholesterol in different matrices			
Sample matrix	Mean recovery %	Sample matrix	Mean recovery %
Butter	0.98	Sheep milk	0.95
Lard	0.98	Meat	1.03
Cow Milk	0.96	Fish	1.05
Tallow	0.97		
Mean	0.99		
Standard deviation	0.040	Relative St. deviation	0.0404

At this point we can calculate the Uncertainty budget can be calculated:

Parameter	Values	Standard uncertainty	Relative standard uncertainty
Method recovery R_m	1.0	0.016	0.016
Matrix effect R_s	1.0	0.040	0.040
Precision	1.0	–	0.012
Combined standard uncertainty			0.045
Expanded u confidence uncertainty (relative) 95%, K = 2			0.089

5.3 Example 2

In this example the estimation of MU from within-laboratory data on bias and precision is reported.

A laboratory has validated a method for the determination of iron in brown bread and has applied the method on samples submitted by customers over a period of six months. To validate the method a certified reference material (BCR-191 certified for iron 40.7 **± 2.3 mg/kg** – 95% confidence range) was used and spiked samples routinely used for within-batch QC were analyzed twice a month.

Now the laboratory is ready to calculate uncertainty not only from validated data but also from the data generated during the six months of analyses of customer samples.

At this point laboratory has data for seven replicate of CRM, obtained during the validation process and 12 data form the analysis of sample from the customer. The analytical process is in control as pointed out by the control charts.

Principal data are reported in the following table

Validation data	Mean	Standard deviation	Relative standard deviation (RSD)
CRM (7 replicate)	35.5 mg/kg	s_{crm} = 2.3 mg/kg	7.0%
Quality control data (12 replicate)	31.2 mg/kg	s_R = 3.3 mg/kg	10.6%
Mean of all results	(35.5 + 31.2)/2 = 33.4 mg/kg		
Mean recovery	82.1%		
CRM	**40.7 mg/kg**	**2.3 mg/kg**	

From the data is evident that the mean obtained under intra-laboratory reproducibility condition are less precise than those obtained during the validation repeatability conditions. However, % recovery and precision data were considered acceptable.

It was considered important to estimate the MU associated with results close to 40 ppm, since it is close to the CRM.

5.3.1 Estimation of Measurement Uncertainty

The mean of all results reported on the third row of the table is the best estimate of the batch-to-batch result. This value is the average (validation + QC replicate measurements) of all observed result:

$$\mathbf{y = 33.4\,mg\,/\,kg;\ with\ a\ mean\ recovery\ of\ 82.1\%}$$

The standard deviation of the QC results best reflects the imprecision of results produced under normal test conditions: $\mathbf{s_L = s_R = 3.3}$ **mg/kg** and the standard deviation of the certified reference standard is 2.3/2 = 1.15. since 2.3 is the expanded uncertainty (95% confidence range) stated for the CRM.

5.3.2 *Estimation of Bias and the Uncertainty of Bias*

To calculate the bias, we must take in account the value of the CRM (40.7 mg/kg) and the value obtained from the QC data (31.2 mg/kg).

The uncertainty of bias is calculated from the following equation:

$$u_{bias} = \sqrt{u\left(\frac{s_r}{\sqrt{n}}\right)^2 + u(s_{CRM})^2}$$

Example data shows that n = 12, so

$$u_{bias} = \sqrt{\left(\frac{3.3}{\sqrt{12}}\right)^2 + (1.15)^2} = 2.23\ mg/kg$$

$$\boldsymbol{Bias(b) = 33.4 - 40.7 = -7.3\,mg/kg}$$

From the t tables (critical values for two-tailed student t-tests), **t (0.05, 12) = 2.201.**

Since −7.3 > 2.201 × 2.3 = 5.06 we can conclude that bias is significant. Consequently, results should be corrected for bias. Below the calculation made with correction of bias or not.

5.3.3 *Result Corrected for Bias*

The correction is very simply since it is sufficient to add the bias to the measured value, assuming the bias to be constant over a narrow concentration range. There is uncertainty associated with this correction, and it is therefore necessary to include the uncertainty of bias in the combined standard uncertainty, irrespective of whether or not results are corrected for bias.

Assuming a raw test result = 32 mg/kg the corrected result would be equal to 39.3 mg/kg (32 + 7.3 mg/kg).

The combined standard uncertainty, can be calculated from the equation

$$\boldsymbol{u_c(y) = \sqrt{s_{bias}^2 + s_r^2} = \sqrt{2.3^2 + 3.3^2} = 4.02\,mg/L}$$

$$U_c = \textbf{\textit{expanded uncertainty}} = 2 \times 4.02 = 8.04$$

By adding the expanded uncertainty, the result would be reported as 49.3 ± 8.04.

5.3.4 *Result Not Corrected for Bias*

The result not corrected for bias is reported by calculating the expanded uncertainty that is:

$$U = 2u_c(y) + b = 16.08 + 7.3 = 23.4mg / kg$$

The laboratory will report the result as 32 ± 23.4 mg/kg.

When the expanded uncertainty is enlarged in this manner to account for uncorrected significant bias, the reported result encompasses a wide 95% confidence range that is wider than justified on the lower side of the measured value, since the uncertainty allows for both negative and positive bias although only negative bias is present.

It si evident that the results corrected for bias is a better estimate of the true results.

5.4 Other Approaches to Estimate MU: The Horwitz Equation

The Horwitz equation is an empirical parameter used as reference value for laboratories quality control activities and proficiency testing programs. (Horwitz W., Kamps L.R., Boyer K.W., J. Ass. Off. Anal. Chem (1980) 63:1344). The original equation was:

$$RSD_r(\%) = 2^{(1-0.5logC)} \tag{5.1}$$

Where C, is the concentration of analyte expressed as dimensionless mass fraction (numerator and denominator have the same units) and RSD_R is the reproducibility standard deviation.

An estimate of repeatability, expressed as relative standard deviation $RSD_{r,}$ was also reported:

$$\mathrm{RSDr} = 0.67 \times RSD_{\mathrm{R}} \tag{5.2}$$

Alternative expressions for the Horwitz equation are reported in the following table:

Expression, units	Mathematical expression
Standard deviation s_R (m)	$0.02C^{0.85}$
Relative standard deviation RSD_R,% (m/m)	$2\ C^{-015}$
Coefficient od variation CV, fraction (m/m)	$0.02C^{-015}$

Where CV = RSD_R (%)/ 100 = s_R / x_C.

s_R = standard deviation among laboratories and x_C the mean or average concentration expressed as mass fraction.

The Horwitz equation has been shown to overestimate the value of the standard deviation at very low concentrations and at very high concentrations. Therefore, in 1999 Michael Thompson transformed the equation into the equivalent form:

$$RSD_R(\%) = 2C^{-015} \tag{5.3}$$

or as reproducibility standard deviation. The following functions are recommended:

Reproducibility standard deviation	Concentration
$S_R = 0.02\ C^{0.85}$	$1.2\times10^{-7} \leq C \leq 0.138$
$S_R = 0.22\ C$	$C < 1.2 \times 10^{-7}$
$S_R = 0.01\ C^{0.5}$	$C > C\ 0.138$

In practice, for concentrations below 120 mg/kg the Horwitz equation would provide values excessively high, so the uncertainty is calculated according to Thompson's correction, that is:

$$\sigma_H = 0.22 \times C \quad \text{where}\ C\ \text{is the effective concentration}$$

For concentrations above 138 g/kg, the Horwitz equation for calculating the standard deviation of reproducibility becomes:

$$\sigma_H = 0.01 \times C^{0.5}$$

The approach described should still be considered with caution and should be confirmed with tests on reference materials (possibly certified) or through participation in inter-laboratory circuits.

In any case, this relationship must be used with caution for the evaluation of measurement uncertainty, even if not it can be neglected, as the equation is used in some cases by the legislator in the field of food analysis.

This criterion is in fact useful, in the initial approach phase, in assessing uncertainty. In fact, in the event that repeatability data of the method in question are not known, it can be used by the operator in the application phases of a standardized method and in the verification of its performances, or in the design phases of an internal method. The laboratory can, in fact, take as a reference the "target" reproducibility provided by the Horwitz equation, and to be reported in terms of repeatability, using the ratio suggested by the author and also required in some cases by the legislator.

Example of calculation of the Horwitz equation as:

$$_{SR} \text{ or } RSD\% \text{ when the concentration } (\mathrm{C}) \text{is} 1\ \mu g/g\left(\mathrm{C}=1 \text{ ppm}=10^{-6}\right):$$

$$_{SR}=0.02\times \mathrm{C}^{0.85}=0.02\times\left(10^{-6}\right)^{0.85}=0.02\times 7.998\times 10^{-6} \text{ or in}$$

$$RSD_{\mathrm{R}}\%={}_{SR}/\mathrm{C}=0.16\times 10^{-6}=16\%$$

A useful table for calculation of measurement uncertainty by the Horwitz relationship can be found below:

C m/m	1.0	0.1	0.0	1×10^{-3}	1×10^{-4}	1×10^{-5}	1×10^{-6}	1×10^{-7}	1×10^{-8}	1×10^{-9}	1×10^{-10}
RSD%	2.0	2.83	4.0	5.66	8.00	11.31	16.00	22.63	32.00	45.25	64.00

Where the RSD% is the percent coefficient of variation of reproducibility and C the concentration expressed as mass fraction, i.e. with the same unit of measurement in the numerator and denominator.

The equation allows to estimate a priori the measurement uncertainty regardless the analyte, method or matrix. The estimated value can be used to evaluate the plausibility of the experimental uncertainty.

If the criteria are met, then the standard deviation of reproducibility calculated according to Horwitz provides an estimate of the laboratory's compound uncertainty, while the expanded uncertainty is equal to:

$$U_e = 2RSD_R$$

where 2 is the usual coverage factor with a probability of 95%.

Therefore, if u.m. is the unit of measurement of the data (Y), the final result can be expressed as follows:

$$\mathrm{Y}\pm 2\ RSD_{\mathrm{R}}\ \text{u m}$$

5.5 HORRAT Value

HORRAT value is the ratio of the reproducibility relative standard deviation, expressed as a percent (RSDR, %) to the predicted reproducibility relative standard deviation, expressed as a percent (PRSDR, %), i.e.,

$$HORRAT = \frac{RSD_R,\%}{PRSD_R,\%} \tag{5.4}$$

Where:

$$PRSD_R\ \% = 2C^{-0.150} \tag{5.5}$$

C = the estimated mean concentration and

$$RSD,\% = \frac{100S_R}{X_{average}} \tag{5.6}$$

Where S_R is the reproducibility standard deviation.

If the concentration C is expressed as trace or mass fraction by power of ten, $C = x_{av}10^{-k}$

K is the exponent. For instance, if k is equal to 6 for part per million(ppm).

When the values of $PRSD_R$, % and RSD_R, % are substituted in the HORRAT relationship by a simply algebraic modification, the equivalent expression can be obtained:

$$HORRAT\_1 = \frac{S_R}{0.02\ x_{av}^{0.8495}} \tag{5.7}$$

This relationship was reported by F.D. McClure and Jung-Keun Lee in the Journal of AOAC International Vol. 86, No. 5, 2003 pp. 1056 in the attempt to calculate the HORRAT value when the analyte concentration is expressed in trace amount, e.g. µg/g (ppm).

In fact, the relationship

$$PRSD_R,\% = 2C^{-0.150}$$

Can be written in the case of trace amount expressed in ppm (10^{-6}) as

$$PRSD_R,\% = 2\left[x_{ay}(10^{-6}\right]^{-0.150}$$

and the HORRAT value can be expressed as

$$HORRAT = \frac{100RSD_R}{2\left[x_{av}\left(10^{-6}\right)\right]^{-0.150}}$$

This is the correct expression to calculate the HORRAT value when the concentration is expressed as usual. However, in the paper mentioned above an interesting table reporting the conversion factors to be used in usual HORRAT calculation (when concentrations are expressed in µg/g (ppm). Below some of these value for 1, 0.1, 0.01, and 0.001 ppm.

Part per million (µg/g)	Conversion factors to be multiplied
1	0.12503
0.1	0.08841
0.01	0.06252
0.001 (1 ppb)	0.04421

HORRAT values between 0.5 to 1.5 may be taken to indicate that the performance value for the method corresponds to historical performance. The limits for performance acceptability are between 0.5 and 2.

The HORRAT can be used as a guide to determine the acceptability of the precision of a method.

The HORRAT is applicable to most chemical methods. HORRAT is not applicable to physical properties (viscosity, RI, density, pH, absorbance, etc.) and empirical methods [e.g., fiber, enzymes, moisture, methods with indefinite analytes (e.g., polymers) and "quality" measurements, e.g., drained weight]. Deviations may also occur at both extremes of the concentration scale (near 100% and .10–8). In areas where there is a question if the HORRAT is applicable, the General Referee will be the determining judge.

The following guidelines should be used to evaluate the assay precision:

HORRAT ≤0.5. Method reproducibility may be in question due to lack of study independence, unreported averaging, or consultations.

0.5 < HORRAT ≤ 1.5. – Method reproducibility as normally would be expected.

HORRAT >1.5. Method reproducibility higher than normally expected: the Study Director should critically look into possible reasons for a .high. HORRAT (e.g., were test samples sufficiently homogeneous, indefinite analyte or property?), and discuss this in the collaborative study report.

HORRAT >2.0. Method reproducibility is problematic.

A high HORRAT may result in rejection of a method because it may indicate unacceptable weaknesses in the method or the study. Some organizations may use information about the HORRAT as a criterion but not to accept the method for official purposes (e.g., this is currently the case in the EU for aflatoxin methods for food analysis, where only methods officially allowed are those with HORRATs ≤2).

Further reading • Measurement uncertainty revisited:

Alternative approaches to uncertainty evaluation, Eurolab Technical Report 1/2007, 2007 (available at www.eurolab.org).

- NORDTEST Report TR 537, Handbook for calculation of measurement uncertainty in environmental laboratories (available from www.nordtest.info).
- ISO 21748 Guidance for the use of repeatability, reproducibility and trueness estimates in measurement uncertainty evaluation.
- ISO 11352 Water quality -- Estimation of measurement uncertainty based on validation and quality control data • B. Magnusson, S. L. R. Ellison, Treatment of uncorrected measurement bias in uncertainty estimation for chemical measurements, Anal. Bioanal. Chem., 390, 201–213, 2008. • G. E. O'Donnell, D. Bryn Hibbert, Treatment of bias in estimating measurement uncertainty, Analyst, 130, 721–729, 2005.

Chapter 6
Control Charts and Process Capability

The purpose of an analysis is to produce comparable analytical results not only in the short term but also in the long term. The basis of quality control are the quality targets defined as the maximum possible error of imprecision and inaccuracy (absolute and relative error). Adherence to these targets during routine analysis must be monitored during the systematic repetition of control analyzes.

In this chapter, the key performance characteristics of analytical methods, the process capability concept and the use of control chart to control its performance characteristics are discussed.

6.1 Control Charts

Internal quality control at the chemical analytical laboratory involves a continuous evaluation of all the routine procedures to monitor the validity of the test results. It is then very important that a laboratory establishes a quality control program where control charts is the central tool to verify if the analytical process is in statistical control. It is well known that routine analyses are prone to error (usually increased bias, but sometimes increased variability) and a control chart can help in identifying errors or deviation (variability) in the analytical process.

A control chart is a graphical plot of test results with respect to time or sequence of measurements, with limits drawn within which results are expected to lie when the analytical scheme is in a state of "statistical control." A procedure i s in statistical control when results consistently fall within established control limits. In addition to identifying results that are out of control, a control chart will disclose trends and cycles, and thus is a tool to provide real-time analysis of data and information u p o n which appropriate corrective action can be based.

S. Mannino, *Guidelines for Laboratory Quality Managers*, Integrating Food Science and Engineering Knowledge Into the Food Chain 14,
https://doi.org/10.1007/978-3-031-11724-4_6

6.2 Construction of a Control Chart

Shewhart or Levey-Jennings control charts are the easiest to construct, use, and interpret. Probably their most useful characteristic is the identification of individual outlying results. To set up a control chart, measurements must be gathered when the process is in control. This means that the analyst must be familiar with the method, he must have explored the various sources of error (for example, sampling procedures, instrumental techniques, and analyte separation) and must have performed sufficient method validation to be certain that the results are acceptable. The following are reasonable steps to be followed in constructing a control chart:

- Identification of the essential elements that compose the chart
- Goal of the measurement
- Selection of the characteristics to be measured
- Decision on how, where, and when the parameter should be measured
- Definition of the sampling procedure
- Obtain reliable estimates of the process mean (if not otherwise known) and of the long-term standard deviation

Control charts help to answer to the question: Is the process in statistical control? Figure 6.1 shows a standard form for a variables control chart with the average (A), action (AL) and warning (WL) limits.

The process average ca be a single observation per run or means of QC results per run.

How to set the control limits for X-Chart (Shewhart or Levis Jennings control charts) The most common method to set limits and central line in a X-charts is to refer to the performance of the analytical method or on the performance of a control sample analyzed for a sufficient long period. The central line will be set at the mean

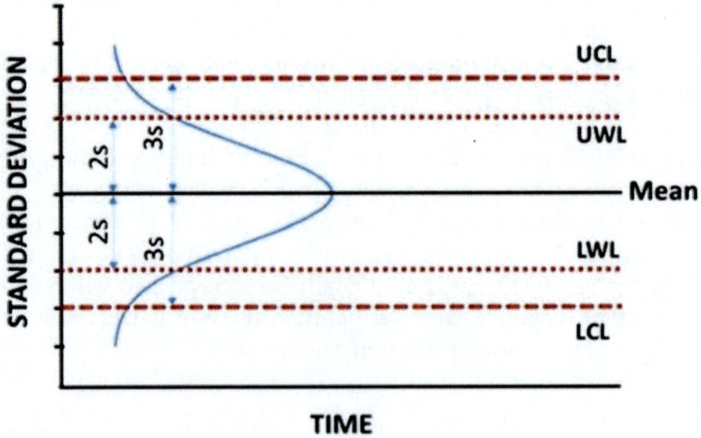

Fig. 6.1 Representation of a Shewart control chart

value of the control sample or in case of well characterized material or reference material the nominal value will be the central line.

Control limits are used as criteria for action or for judging whether a set of data does or does not indicate lack of control. The Warning limits correspond to average ± 2 standard deviations whereas action limits are set at average ± 3 standard deviations. There is only a 5% chance that a result will exceed the warning control limits based solely on random- error considerations, and only a 0.3% chance that a result will exceed the action limit. Exceeding the control limits indicates that precision has worsened or that systematic error may be present. The greater the deviation from the mean, the more likely that systematic error is present.

A process that is "under control" will show a random distribution about the centerline. Any trends or patterns should be investigated for root cause.

Shewhart control charts are the easiest to construct, use, and interpret. To construct a control chart:

- Collect 20–25 observations when the process is control.
- Data must be kept in the exact sequence as they were generated
- Calculate the average and standard deviation
- Calculate the upper and lower control limit by using the standard deviation

As already stated, it should be noted that:

1. 5% of the values may fall outside of the 2σ warning limits
2. 0.3% of the values may fall outside of the 3σ action limits
3. Values falling outside of the action limits should be investigated (Fig. 6.2)

6.3 Type of Control Charts: Average, Range and Standard Deviation Control Charts

Usually, laboratories do not use control charts based on single value but use pair of control chart based on both X-bar and R (range) or X-bar and s control charts. This pair of control charts are used with processes that have a subgroup size of two or more. The X-bar chart shows how the mean or average changes over time and the R chart shows how the range of the subgroups changes over time. X-bar and R charts are typically used when the subgroup size falls between two and ten, and X-bar and s charts when subgroups are eleven or more.

However, before calculating control limits laboratory should collect as many subgroups as possible. With smaller amounts of data, the X-bar and R chart may not represent variability of the entire system. The more subgroups you use in control limit calculations, the more reliable the analysis. Typically, twenty to twenty-five subgroups will be used in control limit calculations. X-bar and R charts are very useful to analyze the results of process improvements. This way to operate has the advantage to use actual test samples and to overcome the problem of lack of reference standards. To examine acceptability, the mean of duplicates and the difference

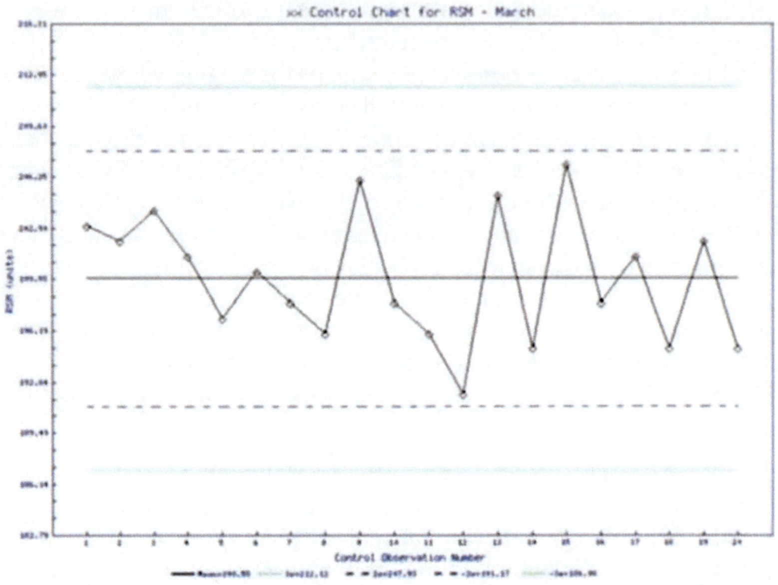

Fig. 6.2 Shewhart Control Chart Form calculated with the calculator reported in https://tools.westgard.com/qccalculator.html. Average 199.55 standard deviation 4.19

between the duplicates are taken into account. A typical example of X and Range charts is reported below in Fig. 6.3.

For R-chart it should be considered that the upper limit is always positive. However, control limits are, as usual, calculated from a standard deviation based on method performance. When measurements are based on duplicates, triplicate and so on factors reported on Table 6.1 can be used for calculating the central line and the control limits.

The mean range is given by:

$$Mean\,range = \frac{\Sigma(\max - \min)}{n}$$

Where n = number of sample

Furthermore, an important advantage of the R-chart is that no special check sample is needed. It is especially useful when standard materials are difficult to obtain; however, the duplicate determinations must be completely independent, that is, not performed simultaneously.

Below an example of Average and Range charts based on duplicate. The procedure is illustrated by the following example based on 10 duplicate analyses.

Duplicate	1	2	3	4	5	6	7	8	9	10
	21.0	24.0	20.0	24.0	19.0	23.0	22.0	19.0	24.0	25.0

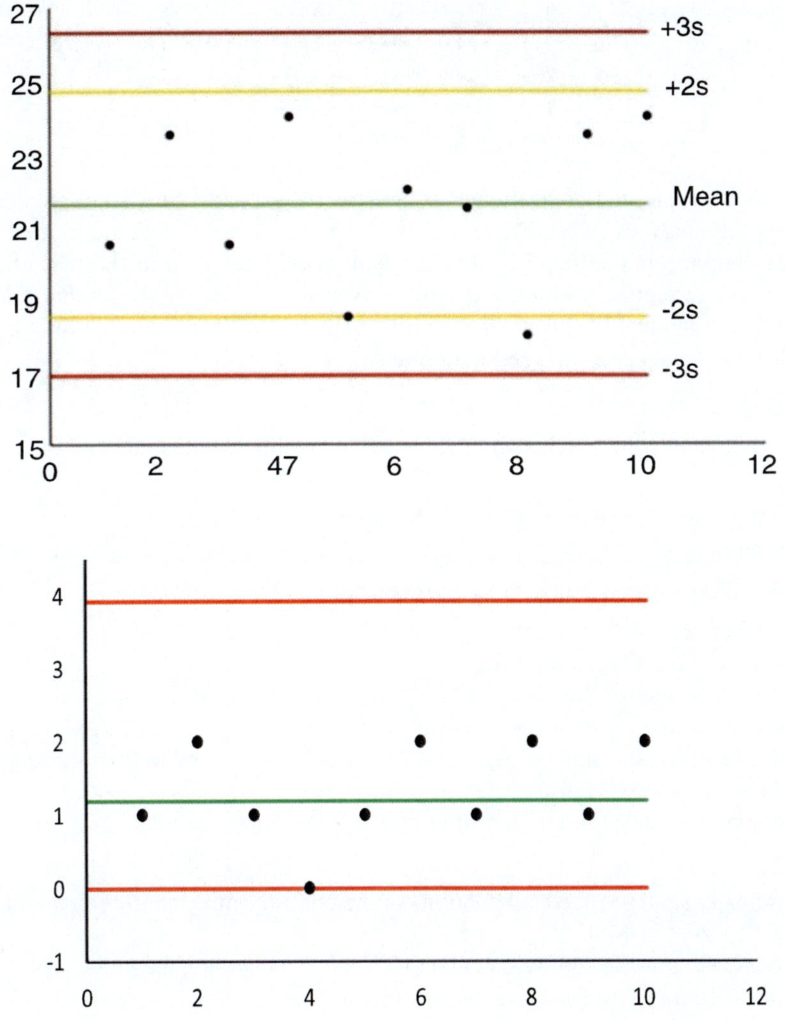

Fig. 6.3 An X-bar and R control chart

Table 6.1 Table of control chart constants

	For sigma estimate	Average and R charts			Average and S charts		
n	d_2	A_2	D_3	D_4	A_3	B_3	B_4
2	1.128	1.880	–	3.267	2.659	–	3.267
3	1.693	1.023	–	2.574	1.954	–	2.568
4	2.059	0.729	–	2.282	1.628	–	2.266
5	2.326	0.577	–	2.114	1.427	–	2.089

Duplicate	1	2	3	4	5	6	7	8	9	10
	20.0	23.0	21.0	24.0	18.0	21.0	21.0	17.0	23.0	23.0
Mean	20.5	23.5	20.5	24.0	18.5	22.0	21.5	18.0	23.5	24.0
Range	1	1	1	0	1	2	1	2	1	2
		Means average = 21.5 and range average = 1.2								

In this example, the control chart is constructed by analyzing 10 duplicate test samples (20 analyses). The differences between the paired results are plotted with zero as the expected value. The second result is subtracted from the first and the differences are plotted with due regard for their sign. The average standard deviation, SD, is calculated using the following equation, where n is the number of test samples that have been analyzed in duplicate:

$$R_{av} = \Sigma \frac{R}{n}$$

Where R_{av} is the Range average and n the number of duplicate.

Procedure:

1. Compute the average and range for each subgroup
2. Compute the grand average X
3. Compute the average range R_{av}
4. The central line for the X –chart is X
5. The central line for the range chart is R_{av}
6. From Table 6.1 the value of A_2, D_3 and D_4 which correspond to the subgroup size (2 in our case) can be found
7. Compute the lower and higher limit by the following equations:

Calculations:

As already said, to speed calculations of the control limits, control charts factors that are reported in Table 6.1 can be used.

In our case, for average we calculate UCL e LCL by using the value reported in the Table 6.1 and using for average the relationship:

$$UCLx = M + A_2\,\Re = 21.5 + 1.88 \times 1.2 = 23.75$$

$$LCLx = M - A_2\,\Re = 21.5 - 1.88 \times 1.2 = 19.0$$

In synthesis, a quality control program based on control charts requires:

(a) Type of control chart
(b) Type of quality control sample
(c) Control limits: warning and action
(d) Frequency of control

Using average and s charts have the advantage of:

- separate the variations in the mean from those in the standard deviation;
- the map s is better than R because for numerous subgroups ($n \geq 5$) the range discards a lot of information, as only the minimum and maximum values of the subgroups are used to estimate the variability
- are efficient in identifying sudden jumps in the average.

An example will illustrate practical application of average, range and SD control charts. Imagine we want control a liquid chromatographic method used for monitoring trace amount of some impurity present in food. For this purpose, one analyst runs five times a day for 20 days (four working weeks) a reference standard material. Data are presented in table below:

	Measured value							
Samples	1.0	2.0	3.0	4.0	5.0	Average	Range	STDEV
1	48.3	47.9	47.9	48.2	48.6	48.18	0.7	0.295
2	48.1	48.8	49.3	47.6	48.6	48.48	1.7	0.653
3	47.3	48.2	48.1	48.3	48.0	47.98	1.0	0.396
4	48.9	48.2	48.2	47.6	48.5	48.28	1.3	0.476
5	49.4	47.8	47.7	48.5	48.8	48.44	1.7	0.709
6	48.7	48.0	47.9	47.5	48.4	48.10	1.2	0.464
7	48.5	49.1	48.0	47.3	48.0	48.18	1.8	0.669
8	48.1	47.3	47.5	47.0	48.0	47.58	1.1	0.466
9	48.2	48.4	47.3	48.0	47.6	47.90	1.1	0.447
10	47.5	48.4	47.5	47.8	48.5	47.94	1.0	0.483
11	47.8	47.4	47.6	47.8	49.0	47.92	1.6	0.626
12	47.5	47.0	48.1	47.7	48.4	47.74	1.4	0.541
13	48.8	49.0	44.0	48.2	48.5	47.70	5.0	2.090
14	47.9	47.7	47.3	48.3	47.6	47.76	1.0	0.371
15	47.1	47.2	47.8	48.5	48.5	47.82	1.4	0.676
16	48.1	48.4	47.7	47.9	47.8	47.98	0.7	0.277
17	48.2	48.5	47.5	47.8	48.6	48.12	1.1	0.466
18	47.8	49.0	47.3	52.8	48.4	49.06	5.5	2.186
19	48.4	48.6	47.1	47.8	48.9	48.16	1.8	0.716
20	47.7	48.3	48.7	48.5	49.2	48.48	1.5	0.550
	Averages					48.09	1.7	0.678

The data we will use are shown in the table. We have 20 subgroups, each containing 5 observations or results. So, the subgroup size is constant for each of the 20 subgroups. The subgroup average, range and standard deviation can be calculated.

The lower (LCL) and upper (UCL) three-sigma (3σ) control limits for the X-bar chart (based on 20 sets of data) are 50.12 and 46.05, respectively, as shown by the green horizontal lines in Fig. 6.4.

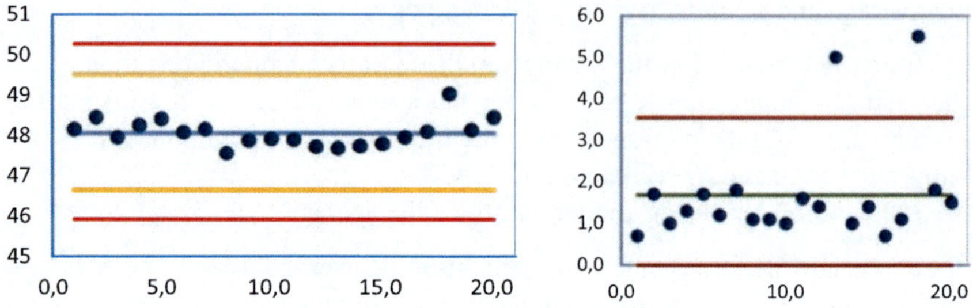

Fig. 6.4 Average and range charts of the above example

The upper control limit for the R chart (UCLR) is 3.59 (1.7 × 2.114) and is shown by a blue line in Fig. 6.4. The lower control limit for the R chart (LCLR) is equal to zero.

6.4 Quality Control Samples

To build a control chart, we need the use of control samples that ideally should be run through the whole analytical procedure. The control sample should be characterized by availability in sufficient amount, stability, and similarity to test samples. They usually are: Certified Reference Material (CRM); Standard solution or in-house control material; Blank samples, Test samples. The results from repeated measurement of these standard materials allows to calculate mean and standard deviation and consequently upper and lower limits. The graph of these parameters (see Fig. 6.1) gives immediate information on both systematic and random effects and allows laboratory to calculate the measurement uncertainty of their results. Repeated determinations in each analytical run give a possibility of using the standard deviation (or range) to estimate of the within lab repeatability of the measurement.

Certified Reference material is an ideal material to construct a control chart. However, often they are not available and are very expansive. CRMs can be purchased ready for use or with a procedure for preparation. A typical certificate is reported at the end of this chapter. It is important to follow the instruction on the method of preparation and conservation before and after use, reported in the enclosed certificate. Standard solutions, in-house or reference materials similarly to CRM, will also give an indication of some of the systematic or random effects. They are prepared in lab but standard solutions can be bought by a qualified supplier and used as received or after preparation. It is important that these solutions are different from the one used for calibration purpose. Finally, stable control sample with known concentration can be used as control sample.

6.5 Guidelines on Interpretation of Control Charts

There are a number of commonly used indicators for interpreting control charts to indicate when the measurement process is out of control. As reported in Fig. 6.1 the Shewhart control chart can be divided into several zones. The following are rules of the thumb that typically apply to control charts to determine if the process is out of control.

- If one point is outside the three-standard-deviation limit
- If two out of three consecutive points are outside the two-standard-deviation limit
- If four out of five consecutive points are outside the two-standard-deviation limit
- If eight consecutive points are in area one standard deviation
- A series of seven or more points above or below the mean, or an increasing or decreasing trend.

However, in the last few years the following six Westgard rules become very popular among medical and biological laboratories.

1	1_{2s}	One measurement exceeds 2 standard deviations either above or below the mean of the reference range.
2	1_{3s}	One measurement exceeds 3 standard deviation either above or below the mean of the reference range.
3	2_{2s}	Two consecutive measurements exceed 2 standard deviations of the reference range, and on the same side of the mean.
4	R_{4s}	Two measurements in the same run have a 4 standard deviation difference (such as one exceeding 2 standard deviations above the mean, and another exceeding 2 standard deviations below the mean).
5	4_{1s}	Four consecutive measurements exceed 1 standard deviation on the same side of the mean.
6	$10_{\times}$	Ten consecutive measurements are on the same side of the mean

The first two rules, 1_{2s}, 1_{3s} are for suspected imprecision or accuracy

The rules 2_{2s}. 4_{1s} 10_x are used for suspected accuracy

The rule R_{4s} is used for suspected imprecision

The website http://www.westgard.com reports examples and tools of how to select statistical control rules and number of control measurements that are appropriate for a specific laboratory test.

Below a graphical representation of the Westgard rules, where:

Point (1) violates rules 1_{3s}
Point (2) is a warning point for the rule 1_{2s}
Points (2) and (3) violate rule 2_{2s}
Points (5) and (6) violate rule R_{4s}
Points (7), (8), (9) and (10) violate rule 4_{1s}

And finally the ten black points on the right violate the rule 10_x (Fig. 6.5)

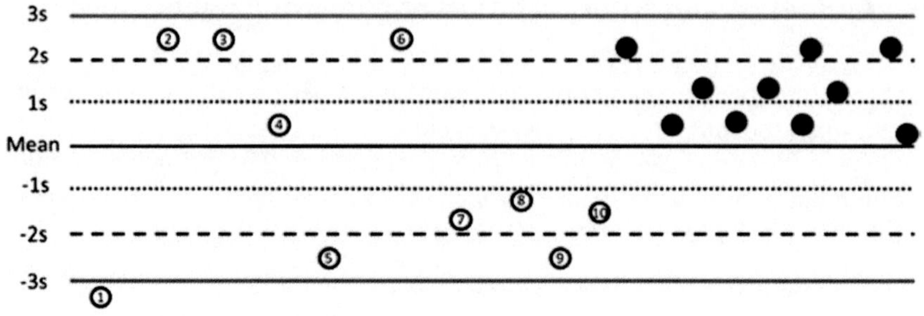

Fig. 6.5 Graphic representation of the Westgard rules

However, not all the Westgard control rules must be used, but they should be selected on the basis of the process capability that we will see at the end of this chapter. In fact, control procedures should be characterized by high probability to detect the error and low probability of false rejection. For instance, by using many control procedures can happen that there is a high probability of detecting the error but at the expense of an increase in false rejection. The power functions (systematic error vs probability of rejection graph) developed by Westgard should be taken in consideration. See the recommended website where some calculation with own data can be done on line.

The interpretation of the R-chart is somewhat similar to the x-chart. Five or more consecutive points above the 50% confidence limit indicate a tendency toward the process being out of control. Any point above the UCLR should be cause for alarm. If R exceed the upper control limit but is below the ULCR limit, accept the measurement but control the next sample. If the last exceed the WCLR, reject all data, take corrective actions to the system. Establish control but before accepting data, at least three consecutive set of data should be in control.

When a decision is reached that a process is out of control, the chart is marked, analyses are stopped, and the cause or causes are investigated. When correction is made, and the system is again in control, charting can be resumed.

6.6 Practical Points in Using a Control Chart

Finally, for a control chart program the first thing to do is the selection of an appropriate control sample (SRM, standard solution and so on) followed by the choice of the control chart type and the establishment of the control frequency. The concentration range is also important especially when environment samples should be controlled. Rules for the evaluation of the control chart should be selected a priori and quality Standard Operation Procedure SOP should describe all operative details including the evaluation and change of the control limits.

Finally, control charts are useful when the laboratory:

(a) use of alternative instrumentation that has been calibrated to provide traceable results;
(b) testing equipment;
(c) replicate tests or calibrations, using the same or different methods;
(d) intra-laboratory comparisons;
(e) participation in proficiency testing
(f) testing of blind sample(s).
(g) and so on

Control limits should be recalculated if the following conditions are all true

- the process has changed
- the cause of the change is known
- the change is expected to last over time
- there are enough data (at least 15 subgroups).

6.7 Process Capability

While analytical method performance characteristics, that have been evaluated during method validation, are essential in the fitness for purpose of a method against a set of requirements, internally defined or regulatory mandated, it is of utmost importance to understand the overall variability in the system observed during routine application, including process and raw material (see Fig. 6.6).

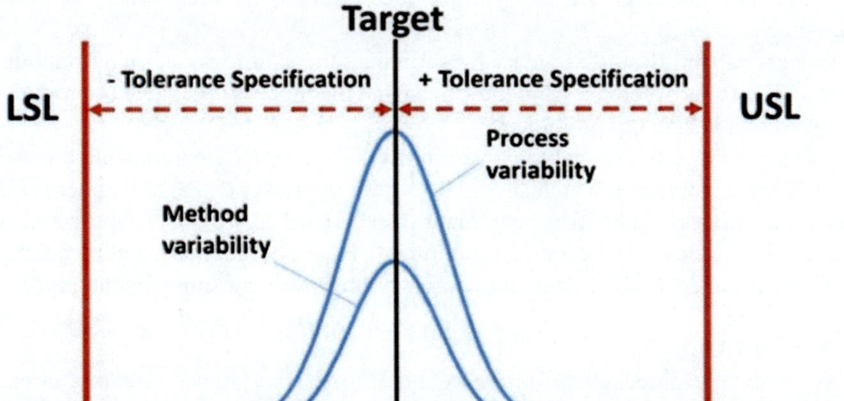

Fig. 6.6 Relationship between specification range, process variability and analytical variability. LSL lower specification limit, USL upper specification limit

The process variability is due to two types of causes:

- common causes (“chance causes) associated with the” natural variability intrinsic to the process.
- special causes (“assignable causes): they are due to occasional events (badly adjusted machinery, operator errors, defective raw material, etc.). The resulting variability is significantly greater than that due to common causes.

A process that operates in the presence of special causes is called out of statistical control, that is that almost all of the results fall between the lower and upper specification limits. When the process is out of control, a certain amount of results falls out of specification.

The main objective of statistical process control is to eliminate the specific causes quickly so that not many nonconforming results are produced. Then bring the process back to statistical control by identifying and removing, one by one, the special causes. For this purpose, as discussed before, control cards are a valid tool for it.

Once special causes have been eliminated and the process is in statistical control, the process capability can be calculated.

In fact, without a proper assessment and deep understanding of the process capability, the risk of wrongly attributing the overall variability on the analytical method that in practice look not fit for purpose while the validation data show that is fit for purpose. Process capability is the ability of a process to produce products or services (analytical results in the lab case) in a state of statistical control (predictable) for a period of time under a given set of conditions.

Process capability is a term that characterizes the tolerance specifications of a result in relation to the central value (systematic error) and variation (standard deviation) of the process. High capacity means that the process can readily produce an analytical result product that is within stated specifications. Conversely, low capacity means that the process is likely to produce products (results) outside the tolerance limits.

Process capability studies are used to compare the natural variation of individual data against stated tolerances and indicates what the process is capable of producing or not what it is actually producing.

As shows Fig. 6.6, the analytical variability should be incorporated into the overall possible variabilities, which should be within the lower (LSL) and upper (USL) limits. The process is said to be centered if $\mu = (USL + LSL)/2$ (that is, if the average falls in the middle of the specification limits). USL – LSL represents allowed tolerance. For instance, CLIA (Clinical Laboratory Improvement Amendments) for cholesterol states a total allowable error of 10% for cholesterol and 15% for cholesterol-LDL

To study the process capability, we refer to the six sigma initiative already described in Chap. 2. In this approach, the standard deviation (SD) represent the process variability in relation to the requirements or specification.

6.8 Capability Indices C_p and C_{pk}

Before describing each feature index, we need to define some terms: the midpoint is the center of the specification limits and is sometimes referred to as the nominal or set point or target value. Tolerance is the distance between the specification limits (Tolerance = USL – LSL).

To have a quantitative method to express the capacity of the process the indices C_p e C_{pk} have been introduced. To calculate these indices, some hypotheses must be satisfied:

- the process is in control
- the data histogram is Gaussian
- the specification limits (LSL and USL) are known

For C_p the following relationship can be used

$$C_p = \frac{USL - LSL}{6\sigma} \tag{6.1}$$

The natural variation of the 6σ process can be interpreted as the range within which almost all measurements of a population are expected to fall (99.7% if Gaussian).

C_p is a measure of the ability of the process to build products that meet the specifications and does not take into account where the mean is with respect to the specifications.

It represents the ratio between the specifications and the process variability that is between the acceptable spread and six time the standard deviation (SD).

$$C_p = \frac{Specification\,range}{6\sigma} \tag{6.2}$$

Assuming the results are normally distributed, C_p can be used as a parameter that indicate the percentage results are outside the specifications. For example, if a parameter in an analytical (or production) process has a C_p of one, it is expected that 0.27 percent of the products will be out of specification at 99% confidence level. The 6 sigma in the C_p formula derives from the fact that, in a normal distribution, 99.73% of the results will be 6 sigma (± 3s).

However, C_p compares the stated tolerance with the natural variability. If the process is centered and Gaussian:

- $C_p > 1$ means that the process is capable of producing almost 100% of acceptable products (less than 3 per thousands of samples are out of specification).
- $C_p = 1$: the process is barely able to produce within the tolerances;
- $C_p < 1$: the process is not capable of producing 100% acceptable results.

The three situations are illustrated in Fig. 6.7 from the left to the right.

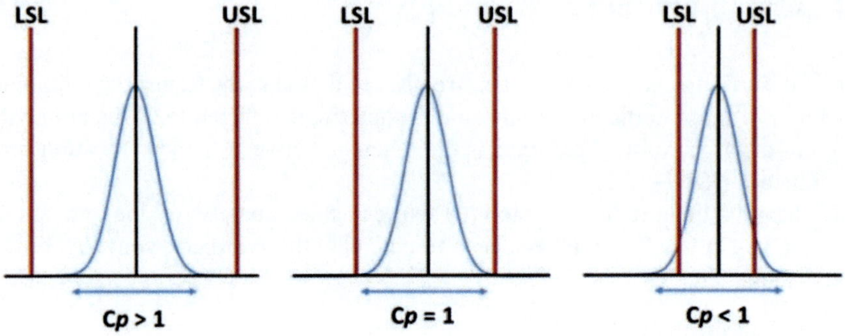

Fig. 6.7 C_p and process

Given a single C_p value, it is immediately known in what relationship the natural variation of the process is with respect to the allowed tolerances. Knowing C_p, the number of analyses expected to fall outside the tolerances, can be calculated.

Practically, C_p measures how well the data fits within the spec limits (USL, LSL). Higher the C_p, smaller the process variability.

However, if the process is not centered, having $C_p > 1$ does not guarantee that the process produces almost all of the products within the specification limits. We only know that it is potentially capable of doing this, but it is not necessarily the case. For this the C_p index was criticized and the C_{pk} index, which takes into account the centering of the process, was introduced. The following relationship are used for C_{pk} calculation

$$C_{pk} = \frac{USL - T}{3s}$$

Or

$$C_{pk} = \frac{LSL - T}{3s}$$

Where T is the target value.

C_{pk} is an index which compares process statistics with allowed tolerances and it takes into account both the variability of the process and the centering.

Figure 6.8 clarifies the difference between C_p and *Cpk*. Gaussian a,b,c, and d, they all have the same C_p but only the Gaussian curve (a) have $C_p = C_{pk}$ since the process is centered.

In Fig. 6.8, shown below, an illustration of a process having with equal C_p but different C_{pk}:

Curve a) $C_p = 2$, $C_{pk} = 2$
Curve b) $C_p = 2$, $C_{pk} = 1$
Curve c) $C_p = 2$, $C_{pk} = 0$
Curve d) $C_p = 2$, $C_{pk} = -1$

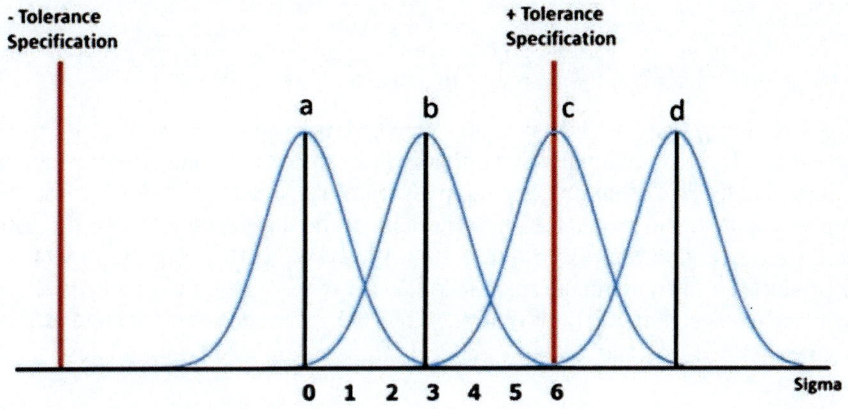

Fig. 6.8 Relation between C_p and C_{pk}

It is clear that the C_p value do not take care about the centering of the process and have the same value independently of the process center line. In general, we are not only interested in the deviations, but also the distance of the mean from the target, which is usually (USL + LSL)/2.

Furthermore, the following relationship between C_p and C_{pk} can be stated by knowing their values:

- $C_{pk} = C_p$ if and only if the process is centered
- $C_{pk} > 1$ and $C_p > 1$: the process is capable and produces within tolerances
- $C_{pk} <1$ and $C_p < 1$: the process is not potentially capable (and does not produce within tolerances)
- $C_{pk} <1$ and $C_p > 1$: the process is potentially capable, not centered and does not produce within the tolerances.

In other words, C_{pk} is a measure of how much near is the average to the nearest limit (higher or lower) of the specification or how centered the data are between the specification limits

6.9 How to Conduct a Functionality Study

The ideal situation is to collect as much data as possible for a sufficiently long period of time. In this way, a very reliable capacity study will be obtained, being based on a sufficiently large sample.

In the process improvement, phases are:

1. collection of process data
2. recording the data on control cards;
3. establishing the control limits;
4. bringing the process under control (i.e. identify and eliminate the determinable causes);

5. calculation of the functionality of the process;
6. if results are not acceptable, process should be improved (reducing the variation by random factors) and then return to phase I.

Typically, a C_{pk} of at least 1.33 is expected, which indicates that the control limit of our test result (for individual values) that is nearest to its tolerance limit is at least 1σ from that limit. The number 1.33 derives from the industry that usually ask their suppliers for a C_{pk} index of 1.33 that means that the difference between the mean and the specification limit is 4σ (since 1.33 is 4/3). With a C_{pk} of 1.33, 99.994% of the products (results) are in the specifications. Likewise, a C_{pk} of 2 is 6σ between the mean and the specification limit (2 is 6/3). The improvement from 1.33 to 2 or more may be justified in some cases to obtain a greater number of results close to the optimal target.

As already stated, C_{pk} measures how close you are to your target and how consistent you are to around your average performance. An analyst may be performing with minimum variation, but he can be, away from his target towards one of the specification limit, which indicates lower C_{pk}, whereas C_p will be high.

6.10 Process Capability Analysis: An Example

Comparison of analytical process data to reference material specifications may indicate the need to adjust the process. For this purpose, we perform the process capability analysis by calculating the usual indices: C_p and C_{pk}.

Example

In a process capability study we found that data are normally distributed and the control X and R control charts indicate that our process is in control. R chart was obtained using subgroup of two.

The following data are available:

Our reference material analyzed for iron content has a value of 40 mg/kg
Upper specification limit: 45.0 mg/kg
Lower specification limit: 35.0 mg/kg
Average of 30 data from the control chart: 38.4
Standard deviation: ±0.165 mg/kg

X bar Upper control limit: 38.89 mg/kg
X bar Lower control limit: 37.90 mg/kg

$$C_p = \frac{45.0 - 35.0}{6 \times 0.165} = 10.1$$

$$C_{pkU} = \frac{38.4 - 35.0}{3 \times 0.165} = 6.86$$

$$C_{pkU} = \frac{38.4 - 45.0}{3 \times 0.165} = 13.3$$

From the above data, it can be seen that our process is quite capable, even though we are off-center by almost 1.6 mg/Kg. Our C_{pkL} is 6.8, it's a remote possibility that the lab makes a wrong analysis.

Let recap: process capability analysis can be done after the lab checked the accuracy and variability of the measurement systems. The analytical process should be checked if unacceptable results are obtained and special causes eliminated. Data should be controlled graphically, by using control charts and by doing the usual descriptive statistics such as average and standard deviation. At this point the capability indices from the available data can be calculated. One observation: C_{pk} cannot be used to show that the process is in control; on the contrary, the calculation of C_{pk} is based on the hypothesis that the process is in control.

6.11 Six Sigma and Process Capability

In the past few years, laboratories have considered the concept of 6-sigma as the quantitative goal for performance acceptability. The fundamental concept of 6-sigma is that 6 standard deviations on each side of the mean of the process should fit within the designated limits, thus ensuring the number of defects (outliers) is very low indeed. The 6-sigma rationale also ensures that any process is relatively robust to unavoidable sources of variation. This is illustrated in Fig. 6.9 which shows the measurement distribution of a process with a 1.5s shift from the mean.

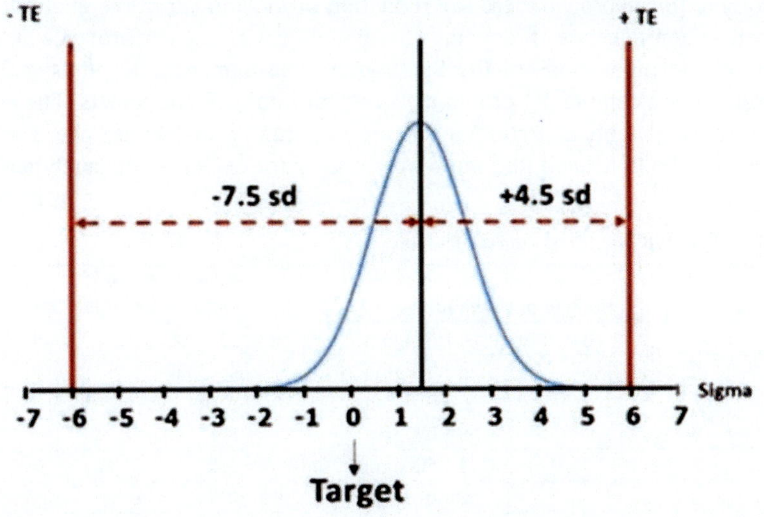

Fig. 6.9 Process with a 1.5 s shift from the target value

Although this was initially a 6 sigma process when there is no bias, the effect of a 1.5 sigma bias reduces the process capability and makes the process equivalent to 4.5 sigma. However, even after a 1.5s shift, the process is still well within the designated limits. This can be considered a good production process if properly controlled but it is always desirable to eliminate the systematic error if possible.

By convention established at Motorola, where the Six Sigma program originated, the Sigma level is adjusted by 1.5 sigma to recognize the tendency of processes to shift over the long term.

In Fig. 6.9, tolerance specifications are replaced by the total error (TE), which are more common for labs.

In other words, a 6-sigma process gives a better assurance that the result is within the desired specification and with a low% defect.

However, the goal of Six Sigma is to design a product so that, for each of its critical responses, the ± TE are 6 standard deviations above or below the mean, respectively (Fig. 6.9). It is assumed that the response has a desired target, which is centered between the specification limit, and that the distribution of the data is normal.

Another way of looking at this is that a 6-sigma process can be monitored with simple QC procedures and any problems that arise would be detected and corrected. As the process capability decreases to 5-sigma or 4 or 3-sigma, the choice of quality control procedures becomes increasingly decisive in detecting any errors. Low capacity processes cannot often even be controlled.

The SEISIGMA system has an accuracy target of TE/6s and in terms of process capability, the 6sigma's combination of precision and accuracy means a C_{pk} of 1.5 (Table 6.2).

By using the capability index approach as described above, laboratories are able to identify assays for which performance can be readily improved and introduce specific control rules to ensure that acceptable performance is maintained. Generally, by adopting this approach there is a reduction of the number of the quality control rules with a consequence of reducing cost and improving the performance of assays.

One of the innovations of the Six Sigma measurement is to adjust measures according to the complexity or number of "opportunities" for defects. The purpose is to make comparable in performance simple operations to complex one. The world complex can be translated into more opportunity for defect. In the analytical field,

Table 6.2 CPK/SIGMA/PPM conversion table

Cpk	Long term sigma limits	PPM defective Outside the limits
0.33	± 1.0 sigma	317.300
0.67	± 2.0 sigma	45,000
1	± 3.0 sigma	2700
1.33	± 4.0 sigma	63
1.50	± 4.50 sigma	3.4
1.67	± 5.0 sigma	0.6
2	± 6.0 sigma	0.002

some method can be very simple other complex and it is important to define the number of opportunity to make an error. One way to calculate or express measures based on defect opportunity is to calculate the so called DPO (Defects for Opportunities).

$$DPO = \frac{Number\ of\ defects}{\neq\ of\ results \neq\ of\ opporunities}$$

For instance, if the number of defects are 10, the number of result 50 and the number of opportunities 4 the DPO will be 0.05 that signify a five percent chance to have a defect. In Six Sigma DPO are often converted in Defects per Millions of Opportunity or DPMO just multiplying DPO by 10^6. This allows to translate in Sigma performance by using a conversion table (see Table 6.2). In our case DPMO are 50,000. That is equivalent to about 3.1 sigma process. The following table lists Defects Per Million Opportunities with the corresponding Sigma Level and C_{pk}. A complete table can be found in internet.

A ± 3 sigma process generate 99.73% of good results. This means that 2700 analyses over 1.000.000 will be outside the limits. This processes are usually not acceptable and the precision should be ameliorated to increase the sigma level.

In a 6 sigma design, there are approximately a 2 in a billion chance of exceeding the limits. However, a centered response is a rare condition, and probabilities increase dramatically as the response drift from the center. In fact, it is common to assume that a 6 sigma design will have a 3.4 in a million chance of exceeding the specification limits, not 2 in a billion. This is based on the assumption that in a long term the mean of the response will vary by 1.5 times its standard deviation (Fig. 6.8).

In term of capability index, a 6 sigma performance will have a $C_p = 2$ and $C_{pk} = 1.5$:

$$Cp = \frac{USL - LSL}{6\sigma} = \frac{12s}{6s}$$

In term of C_{pk}, if we use a shift in standard deviation of 1.5 we have:

$$Cpk = \frac{USL - X}{3\sigma}; \frac{X - LSL}{3\sigma} = \frac{6\sigma - 1.5\sigma}{3\sigma}; \frac{1.5\sigma - (-6\sigma)}{3\sigma} = 1.5$$

The following table will give some relationship between capability and 6 sigma process (Table 6.3):

Table 6.3 C_p, C_{pk} and sigma relationship

Sigma	1	2	3	4	5	**6**
Cpk	−0.16	0.16	0.5	0.83	1.16	**1.5**
Cp	0.33	0.66	1	1.33	1.66	**2**
With a maximum process shift of ± 1.5						

Sigma level is calculated by dividing the process's allowable tolerance (upper specification minus lower specification) by twice the process's sigma value, since the sigma level of a process is normally stated as a ± value.

For instance, for a process having a standard deviation of 0.67 mg/L, the maximum allowable value of 12 mg/L and a minimum allowed value of 8 mg/L. The process sigma level is:

$$Process\,sigma\,level = \pm\frac{Process\,Tolerance}{2\sigma} = \pm\frac{4}{2\sigma} = \pm 3$$

It is obvious that the process is poor and we should improve it by increasing the sigma level. This could be a quality goal. If this is not possible, control procedures must be improved or adapted so that they are capable of detecting errors while keeping false rejects low. For this purpose, Westgard has introduced the so-called Power Function Graphs which show the rejection characteristics of a quality control procedure generally used in laboratories. These quality control procedures are based on the critical systematic error **(ΔSEc)**.

6.12 Process Capability Index C_{pk} and Six-Sigma Metric

Thanks to the work of Westgard et al., the use of the critical systematic error (ΔSE_c) and critical random error (ΔRE_c) in the medical field is now widely used especially for the selection of quality control rules. They have described the following relation between C_p and the critical systematic error (ΔSE_C) assuming that bias is zero.:

$$\mathbf{\Delta SE_C = 3Cp - z}$$

where z is a factor for a one-tailed test of significance (usually set at 1.65 for 95% confidence, assuming a Gaussian distribution).

It has also been shown that when bias is not zero the following equation apply:

$$\Delta SEc = \frac{TE_a - bias}{s} - z \tag{6.1}$$

Where TEa is the allowed total error, bias the observed accuracy, s the standard deviation and z set as usually to equal 1.65 (only 5% of the tail of the normal distribution exceeds the quality requirements. (see the last Gaussian in Fig. 6.10) Furthermore, to express the process capability in term of the six sigma system the following relationship apply:

$$Sigma\ process = \frac{TEa - bias}{s_{meas}}$$

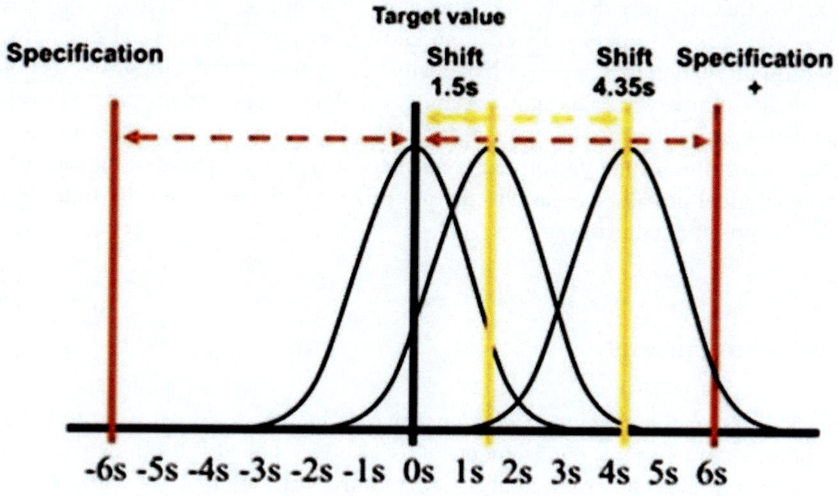

Fig. 6.10 Six sigma process with a shift of 4.35s

And then,

$$\Delta SE_{\text{crit}} = \text{Sigma process} - \text{z}$$

For a 6-sigma process the ΔSE_{crit} will be 4.35 (see Fig. 6.10). This means that it is necessary to implement a control rule that is capable to detect a systematic error of 4.35 times the standard deviation. In this case a single control procedure $1_{3.5s}$ or 1_{3s} with two control samples would be appropriate to detect an error with low probability of false rejection. For a 5-sigma process the systematic shift would be of 3.35 (5–1.65) and it should be detected by using different control rules that can be chosen by using the Westgard power function graphs.

In fact, quality control sample are widely used to monitor the quality of analytical methods but different strategy should be adopted in order to have high probability of detection and low probability of false rejection.

Quality in the laboratory therefore requires the selection of suitable control rules and the definition of the number of necessary controls.

Control rules designated to have low probability for false rejection and high probability error detection can be depicted by using the "power function graph" as described by Westgard (for explanation see the Westgard website (www.westgard.com). In this website, as said before, can be found many interesting tools such as power function graphs, Qc selection grids and recently the "Westgard Sigma Rules TM". Some tool can be used for free.

However, it is important to understand that the ability to detect the error, the control procedure is connected to the process capability. In fact, with a high process capacity, the error that can cause a non-compliant result will be large and therefore it can be more easily detected. On the contrary, when the process capacity decreases, the errors that must be detected become smaller, therefore a better QC procedure that has the ability to detect the error should be selected. However, at present it is very difficult that laboratories reach capacity of 5 or 6 sigma so a lot of work should be done to improve the analytical variability and laboratory organization. However, many laboratories run too many unnecessary quality control samples and then apply the Westgard rules to the performed control chart. The knowledge of the capability of an analytical process can help in the optimization of the process by running the optimal amount of control samples.

6.13 Conclusions

The quality demand on laboratories continue to increase and we need to have in mind the customer or law requirements of the produced results. Running more quality control samples QC or selecting stricter control rules more EQA will not necessarily achieve this end. Processes must be put in place to maximize error detection and improve performance without necessarily increasing laboratory costs.

Chapter 7
Risk Management

7.1 Risk Management Requirements in the New Laboratory Standard ISO17025

Critical components of any high quality testing program in a laboratory are assessment and management of risk. The new ISO 17025:2017 requires the laboratory to carry out actions that address risks and opportunities within the laboratory management program. These actions should be planned to improve the program so that testing can be performed properly, and most importantly can prevent failures in the laboratory. The ISO 17025:2017 prescribes that actions should be taken to address risks and opportunities that are proportional to the potential impact on the laboratory results. That is, the higher the risk and seriousness of the potentially negative impact on analytical data, the more dynamic the actions to prevent or reduce those risks should be. Remember that results are the final product of the laboratory activities. Below the structure of point 8 of the ISO17025:2017 where the management activities are outlined.

	8.1	Options (A&B)
	8.2	Option A – Management System Documentation
	8.3	Option A – Control of Documents
	8.4	Option A – Records
Point 8 Management Requirements	8.5	Option A – Risks & Opportunities
	8.6	Option A – Improvements
	8.7	Option A – Corrective actions
	8.8	Option A – Internal Audits
	8.9	Option A – Management Reviews

S. Mannino, *Guidelines for Laboratory Quality Managers*, Integrating Food Science and Engineering Knowledge Into the Food Chain 14,
https://doi.org/10.1007/978-3-031-11724-4_7

7.2 Addressing Risks

The ISO 73:2010 defines risk as a "combination between the consequences of an event (including the circumstances changes) and the plausibility of occurrence". There is a wide range of risks, from very low risks to very high risks, because there is no risk-free activity. Events with a low level of severity, but with a frequent occurrence involve higher risks, and events with high severity even if they are isolated cases involve very high risks. So, the role of lab managers is to keep the risk at an acceptable level for customers and patients in case of medical laboratories.

A laboratory should identify potential risks encountered in the planned operations and reduce them. Risk management involves decision-making based on the likelihood and severity of potential consequences from failure or error. Sometimes an informed decision can lead to certain risks becoming acceptable, and these risks should be monitored regularly.

7.3 Addressing Opportunities

Seizing opportunities for improvement in the laboratory should lead to better testing quality overall, and encompass a broad range of actions to optimize testing through superior technologies, newly published methods and quality assurance that reduces error in the entire process. In addition, actions to address opportunities can include improved customer management and satisfaction, as well as providing additional value to clients. Preventive action is part of addressing opportunities and often involves creating conditions that make it difficult for errors to occur. Preventative measures related to safety must be included.

7.4 Integrating and Implementing Actions

The risk management plans should be realistic, sustainable and efficient, and possibly with no increase in operational costs. As soon as areas for improvement and potential risks are identified, the laboratory must implement solutions able to prevent or reduce these risks. At the same time, evaluating the likelihood of these risks to materialize can be useful in determining the frequency and timing of these actions.

Implementation of the actions should be based on concrete information, such as audit results, corrective action plans, and review of procedures. The results of these periodical evaluations should be used in adapting, eliminating, or introducing actions that address risks and opportunities. Internal audits can reveal failure or success of actions that address risks and opportunities. Corrective action investigations and root cause analysis can also serve to pinpoint areas that are weak and error prone. In any case, actions to address risks should theoretically and in practice prevent significant impact on laboratory results when successfully implemented.

7.5 Risk Management

The evaluation of the conditions that can lead to the errors' occurrence can be achieved by applying the risk management approach in the laboratory.

According to Guide ISO 73:2010, definition 2.1, "risk management is described by coordinated activities which have as a purpose to direct and control an organization concerning the risk".

In practice, once the risks have been identified, the laboratory must implement continuous monitoring and control processes, in order to be sure that the risk is kept at an acceptable level.

Furthermore, the Guide ISO 73:2010 defines the following concepts:

1. "Risk identification is the process of discovery, recognition and description of the risks. Risks' description is a structural presentation and it contains four elements: sources, events, causes and consequences."
2. "Source of risk is the element that alone or in combination with other elements, has the potential to cause an inherent risk."
3. "The event is the appearance or change of a particular set of circumstances."
4. "The consequences are the effect of an event which affects the objectives".

Depending on the size of the organization the risk management function may range from a single person, a part time risk manager, to a full scale risk management department.

The role of the Risk Management function should include the following:

- *setting policy and strategy for risk management*
- *building a risk aware culture within the organization including appropriate education*
- *establishing internal risk policy and structures for business units*
- *designing and reviewing processes for risk management*
- *coordinating the various functional activities which advise on risk management issues within the organization*
- *developing risk response processes*
- *preparing reports on risk for the laboratory management*

Risk management processes can be customized, expanded, and extended to suit individual organizational needs. Optimally, one person should serve as the risk manager for the lab, even if that person has other responsibilities—as is commonly the case. The risk manager leads the overall management approach, leads the risk management team, reports to management, and chairs risk team meetings. The risk management is usually composed by risk analysts that are person engaged in other activities in the lab. They perform discovery, investigation, and reporting activities that support the risk team's management efforts. The primary tracking and communication tool for your risk management program is the risk register. It can be something as simple as a spreadsheet or a multi-table relational database. It should track at least the following metadata for each risk:

Risk title	A short "name" for the risk
Risk description	Narrative description
Identification code	A unique identifier for every risk
Identification date	An aid for tracking and prioritizing
Identified by	An aid for tracking, assessment, and control
Potential impacts	A narrative description of what could result from the risk
Severity	Commonly "high/medium//low,"
Status	
Owner	The person in charge of this risk.

Like all risk management activities, remember to use it to ensure that the team feels confident in the technical solutions adopted and no major issue was missed.

For more information about medical device risk management system consult the ISO14971:2007 (Medical devices — Application of risk management to medical devices).

7.6 Risk Identification

Risks analysis means to establish the consequences and probabilities when the risks may occur given the presence/absence and efficiency of any control types. The risks' consequences and probabilities are subsequently combined to establish a risk level.

Regardless of the techniques that are used to identify the risks, is necessary to pay attention to the human and organizational factors, besides the hardware and software events.

The identified risks must be related to the organization's objective and it must also be done according to a specific approach or using a certain tool.

Methods of identifying risks may include:

1. Documented methods, for example lists to check and to review historical data.
2. Systematic approach, to identify the risks
3. Reasoning techniques such as FMEA (Failure Mode and Effect Analysis), FTA (Fault Tree Analysis) FRACAS (Failure Reporting, Analysis, and Corrective Action System) or HACCP (Hazard Analysis Critical Control Points).

EP18-A2 proposes two techniques for identifying and control the errors in the laboratory: FMEA and FRACAS, while the hazard analysis is required by the ISO14971:2007.

However, the ultimate goal of risk analysis in food testing laboratories is to create a list of risks with their impact (severity) on the laboratory management process and to determine their occurrence. Once this is done, it is possible to calculate a risk factor that allows us to create an intervention priority scale to mitigate the risk.

The first step is therefore to create a risk assessment model based for example on the Failure Modes and Effects Analysis (FMEA) or hazard analysis (HACCP). The first follows a bottom-up approach while the second follows a top-down approach. However, the FMEA is preferable because *is a technique by which the consequences of an individual fault mode are systematically identified and evaluated.* It starts at a low level of the product or process, working its way up to the effects to the system of subsystems. In a **top-down analysis (like the Hazards Analysis)** it is difficult to identify correctly all the low level causes and sometimes, with complex systems, it is not all that practical.

Let's look at the similarities and differences of the FMEA and the Hazard Analysis with the help of an example concerning balance calibration. Some template similar the one reported below are prepared.

FMA Template

Item	Failure mode	Effect	Cause	Severity	Occurrence	RPN
Item/ component	How can this item fail?	What happen to the system/user				
Balance	Wrong weighing	Wrong results	Calibration plan ineffective			

HACCP Template

Hazard	Hazardous situation	Harm	Cause	Severity	Occurrence	Classification
Mistake in the sample weight	Balance not calibrated	Wrong results	Calibration plan ineffective			

The FMEA starts with the low level components on the left side and progressing to effects, causes, initial evaluation, risk mitigation activities and final evaluation.

Following is a simplistic view of **FMEA** – the balance can **fail** due to wrong calibration caused by an ineffective calibration plan. The **effect** is that the balance can be infective and gives wrong results (**hazardous situation**):

The same case in a top-down approach of **Hazard Analysis** – the **hazard** of weighing the sample can result in a **hazardous situation** where the analyst obtains wrong result. This can be **caused** by the wrong calibration of the balance, because of its ineffective calibration plan.

To summarize, the **main link between the FMEA and the Hazard Analysis** is at the **Cause** level: a certain **failure mode** has the potential to result in a **hazardous situation**; and the **hazard** related to this hazardous situation **is caused by** (among others) that **failure mode**.

7.7 Failure Modes and Effects Analysis (FMEA) for Laboratory

Failure Modes and Effects Analysis (FMEA) is methodology for analyzing potential problems early in the development cycle where it is easier to take actions to overcome these issues. The idea behind this technique is to identify functions whose failure has a very undesirable effect on the safety of a system or can lead to overall poor performance.

Steps for an FMEA analysis can be:

1. Describe the process and its function. Create a block diagram of the process.
2. Identify failure modes – A failure effect is defined as the result of a failure mode on the function of the process. A failure mode in one component can serve as the cause of a failure mode in another component.
3. Describe the effects of those failure modes.
4. Establish a numerical ranking for the severity of the effect. This will help the analyst to determine whether a failure would be a minor nuisance or a catastrophic occurrence to the customer. This enables the analyzer to prioritize the failures and address the real big issues first.
5. Identify preventive measures
6. Determine the likelihood of detection
7. Evaluate each failure mode as per its Risk Priority Number (RPN) which is a mathematical product of the numerical Severity, Probability, and Detection rating. In this manner the team evaluates all the process steps and sub process steps that fall within its scope of work and suggest recommended actions. Thus chances of error are minimized, ensuring patient's safety.

Once we calculate the RPN, recommended action(s) to address potential failures that have a high RPN, should be determined. Responsibilities and the time of completion of these actions should be assigned. After these actions have been taken, the severity, probability and detection and review the revised RPN's, should be re-assessed.

Severity and probability are most important for risk assessment and management:

- Severity in general means: how big is the problem if it occurs?
- Probability means: what is the likelihood that a problem occurs?

Generally, probability include severity and severity factors get higher if the probability of detection is low.

Apart the technique used, while identifying the risk, the following questions should be considered:

What could happen: what might go wrong, or what might prevent the achievement of the relevant goals? What events or occurrences could threaten the intended outcomes?

How could it happen: is the risk likely to occur at all or happen again? If so, what could cause the risk event to recur or contribute to it happening again?
Where could it happen: is the risk likely to occur anywhere or in any environment/place? Or is it a risk that is dependent on the location, physical area or activity?
Why might it happen: what factors would need to be present for the risk to happen or occur again? Understanding why a risk might occur or be repeated is important if the risk is to be managed.
What might be the impact: if the risk were to eventuate, what impact or consequences would or might this have?

7.8 Probability of Occurrence

Probability of occurrence explores the likelihood that an identified risk could occur. Probability of occurrence uses a rating and value scale ranging from Not Present to Almost Certain to Certain. It can be scaled in different ways. Below are reported some common value scales for probability of occurrence.

The first chart scale is very simple and it is arbitrarily scaled to 1–3 with 3 being the highest probability.

1	Unlikely to occur
2	Likely to occur
3	Very likely to occur

The second chart scale is arbitrarily scaled 1–5 but it also includes a percent probability an issue will occur.

7.9 Probability of Occurrence with Standard Linear Scaling

Occurrence value	Probability of occurrence	
Value	Percent	Rating
1	0%	Item or operation not present in the lab/impossible
2	1–10%	Rare
3	10–50%	Possible
4	50–90%	Likely (improbable)
5	90–100%	Almost certain to certain

Another way to deal with the probability of occurrence is to utilize in alternative the following table that can be connected to the Cpk of the process and then to the six sigma methodology (see Chap. 2).

Probable Failure	Failure rate	Cpk	Occurrence rating
Very high (inevitable)	>1 in 2	<0.33	10
	1 in 3	0.33	9
High (often)	1 in 8	0.51	8
	1 in 20	0.67	7
Moderate (occasional)	1 in 80	0.83	6
	1 in 400	1.00	5
	1 in 2.000	1.17	4
Low (isolated)	1 in 15,000	1.33	3
Very low	1 in 150,000	1.50	2
Remote	<1 in 150,000	1.67	1

7.10 Severity

At this point we must rate the different impact on the laboratory activities of each failure mode. This allows to calculate the severity of the consequence by assigning a value. It is possible to use a Standard Linear Scaling or better a Weighted Value scale chart. Each failure mode can have different impact such as to personnel safety, work performance, property damage, and laboratory reputation. These impacts can be associated to a rating. For instance, in the chart below we have an arbitrary scale of 1 to 4, with 4 being the highest severity.

Consequence value			Impacts			
Rating	Linear level	Weighted level	Personnel safety	Work performance	Wrong final results	Lab reputation
No risk	**1**	**1**	No injuries	No delay	No impact	No impact
Minor	**2**	**5**	Minor injuries	Modest delay	Modest	Potential damage
Moderate	**3**	**10**	Moderate to life impacting injuries	Significant delay	Significant	Damage
High	**4**	**20**	Life threatening injuries from single exposure	Major operational disruptions	Wrong results	Loss of confidence

The primary goal of risk rating is to differentiate between laboratory's high-risk activities and low- risk activities. Therefore, it may be necessary to weight your consequences value scale to meet your laboratory's existing priorities and protocols (weighted level are reported in the third column).

For example, using the standard linear scaling, an activity with a certain probability of 4 (likely) with no risk (level 1) would produce an overall risk rating of 4 (1×4). An activity with a probability value of 1 with potentially lethal consequences (4) would also result in a risk rating of 4.

Since any activity with the potential of being lethal would not be considered low risk, regardless of how low the probability, the weighted scale reflects better the severity of potential consequences. The Weighted Consequence Value Scale demonstrates how you might assign consequence values to achieve risk ratings more reflective of the impact of moderate and high consequences.

Let's look back to our previous example. Using this weighted scale, an activity with a certain probability (4) with no risk (1) would still produce an overall risk rating of 4. But, an activity with a rare probability value (1) with potentially lethal consequences (20) would result in an overall risk rating of 20. into risk codes (see below).

7.11 Risk Mitigation

This phase evaluates whether or not the risks as estimated in the previous phase are acceptable. If they are not acceptable, alternatives on how to mitigate risks should be evaluated and implemented. Starting point is the risk factor as calculated before. The risks can be put into three categories as defined in the following table (Table 7.1).

All risks with factors higher than 20 (Code 3 = High risk) must be mitigated, all risks with factors higher than 10 (Code 2 = Medium risk) should be considered for mitigation. The following table can be used to document the mitigation strategy, costs for mitigation and non-mitigation, and the decision whether to mitigate the risk or not. It is important to estimate the cost for mitigation as well as potential losses through non-mitigation. Losses for non-mitigation should include real or direct costs and tangible or indirect costs.

<table>
<tr><td>Person:
Date:</td><td colspan="2">System ID:</td><td colspan="2">Location:</td></tr>
<tr><td>Risk Description/ID</td><td>Mitigatio n Strategy</td><td>Cost of Mitigation</td><td>Cost of Non-mitigation</td><td>Mitigate Yes/No</td></tr>
<tr><td></td><td></td><td></td><td></td><td></td></tr>
</table>

Table 7.1 Definition of risk codes and consequences

Code 1	**Risk factor 10 and lower:**	Routinely accepted, no action taken.
Code 2	**Risk factor 11–20:**	Operation requires written, time limited Waiver, endorsed by management. Mitigation subject to cost/benefit analysis.
Code 3	**Risk factor higher than 20:**	Not accepted. mitigation required. Alternative approaches should be evaluated.

7.11.1 *Table Form for Risk Mitigation*

There are different ways and approaches to mitigate risks. They can range from the process change to a better qualification of the staff passing through a careful validation and monitoring Typically, the most effective ones are also the most expensive and take the longest time.

Once the decision to mitigate has been made and the strategy is identified, a plan should be developed regarding the time schedule and deliverables.

When the plan is in place and the system is running, the effectiveness of the plan should be monitored, reviewed and adjusted if necessary. The monitoring program should also help to identify previously unrecognized hazards. These could have been introduced by changing processes or introducing new technologies.

7.12 Detection Level

Another important parameter to establish in risk analysis is the detection level (D). Detection is an assessment of the likelihood that the Current Controls (design and process) will detect the Cause of the Failure Mode or the Failure Mode itself, thus preventing it from reaching the Customer. Based on the Current Controls, consider the likelihood of Detection using standard tables for guidance. Detection is usually rated on a scale from 1 to 10, where 1 means the control is absolutely certain to detect the problem and 10 means the control is certain not to detect the problem (or no control exists).

7.12.1 *Detection Level Table*

Level	Description
10	Nearly certain that failure won't be detected (p almost 1)
9	Extremely poor chance of detection
8	Poor chance of detection
7	Highly unlikely to be detected before reaching customer
6	Unlikely to be detected before reaching customer
5	Might be detected before reaching the customer
4	Low probability of reaching the customer
3	Low probability of reaching the customer without detection
2	Extremely low probability of reaching the customer without detection
1	Nearly certain to detect before reaching customer

The risk responsible for each cause, should identify current process controls. These are tests, procedures or mechanisms that you now have in place to keep failures from reaching the customer. These controls might prevent the cause from happening, reduce the likelihood that it will happen or detect failure after the cause has already happened but before the customer is affected.

7.13 RPN Calculation

One the severity, the probability of occurrence and the detection method for each failure mode, was set up, we can calculate the RPN. Also preventive measures and the likelihood of detection should be determined. Whit these data we can calculate the Risk Priority Numbers (RPN), which is a mathematical product of the numerical Severity, Probability, and Detection rating:

$$RPN = (\text{Severity}) \times (\text{Probability}) \times (\text{Detection})$$

The numerical value of RPN depends on the scales used for Severity, Probability of occurrence and Detection.

7.14 Examples

Two practical example are reported below. One chosen from a medical lab and the other from a food control laboratory. Both examples use 3 scales (probability of occurrence, severity, and detection level) all based on minimum numerical value of 1 and 10 for maximum. Below the three scales:

	Risk impact	
10	Very high	Impact on safety or compliance with the law
9–8	High	Impact on the customer
7–6	Moderate	Impact on the final results
5-4-3-2	Low	No influence on lab operability
1	None	Unnoticed

Table 2

Occurrence scale table		
Rank	Interpretation	Criteria
10–9	Very high	Many in a day
8–7	High	Many in a week
6-5-4	Moderate	Once a week
3	Low	One a month
2	Very low	Once every quartermaster

Occurrence scale table		
Rank	Interpretation	Criteria
1	Remote	Once ever

Table 3

Detection table		
Rank	Interpretation	Criteria
10	Extremely low	No control available
9	Very low	Control probably will not detect
8–7	Low	
6–5	Moderate	Control is able to detect after reaching next operation
4–3	High	Control is able to detect before ending the operation
2–1	Very high	Control will certainly detect

7.14.1 Medical Example: Prostate Specific Antigen (PSA) Test

This example concerning PSA testing is based on review of laboratory analysis processes. This is a high-volume process with potentially severe outcomes should the process fail. Assume that the topic and selecting the team have been accomplished. Therefore, the example starts with developing the process flowchart.

As a next step the team gathers information about how the process works. In this example the test is ordered, the phlebotomist draws the sample, and then the sample is sent to the laboratory for analysis. The analysis is conducted, and the results are reported to the physician and recorded in the patient's medical record. The failure mode is evaluated, and then the failure mode causes are identified and evaluated. The team assesses the severity, probability & chance of detection and assign numerical value, using the definitions provided by the 3 tables.

Failure mode	Severity	Probability	Detection	RPN
Wrong test ordered	3	7	2	42
Order not received	1	8	3	24
Equipment broken	5	7	4	140
Wrong test tube	8	7	6	336
Instrument not calibrated	4	1	2	8

Each failure mode is evaluated as per its RPN. In this manner the team evaluates all the process steps and sub process steps that fall within its scope of work and suggest recommended actions. Thus chances of error are minimized, ensuring patients safety.

7.14.2 *Example 2: HUMIDITY for Rice and Mill Analysis*

As stated before, each method in the laboratory is a process (the analytical process) that start with the sample receipt to the delivery of the results to the customer.

To show how the FMEA analysis works, let's apply it to a simple method like the "Dry air oven method for flours and ground wheat".

The lab receives the sample that is identified and then submitted to the analytical process. The principle is that the sample is heated under certain conditions of temperature and time, and the loss of weight is used to calculate the moisture content of the sample. The principle of moisture determination by drying to constant weight applies. The analysis is conducted, and the results are recorded and reported to client.

This example is based on review of laboratory analysis processes. The focus was directed at the humidity parameter for rice and mill analysis. This is a critical parameter in many food analyses with potentially severe outcomes should the process fail. In fact, to calculate other parameters such as protein and fat on a dry base an accurate value of the humidity is needed. Assume that the topic and selecting the team have been accomplished. Therefore, the example starts with developing the process flowchart considering the following steps:

1. Weigh base and lid accurately. Recorded in data table. (Ensure that the base and lid have matching numbers)
2. Placed 2–3 g flour in the pan and weigh accurately. Recorded in data Table
3. Placed flour in a fan forced draft oven at 130 °C for 90 min. Distribute samples with consideration for effects of oven position. Ensure metal covers are ajar, to allow water to evaporate.
4. Placed in a desiccator until cool for weighing. Weigh closed dish with dried specimen accurately. Recorded in data table.
5. Repeated oven drying for a further 30 min interval until dried to constant weight. Recorded weights after each drying.
6. The % moisture (wt/wt) are calculated

An example of table for the above example can look like this:

Failure Mode	Severity	Probability	Detection	RPN
Sample identification	7	3	4	84
Sample homogeneity	8	5	3	120
Lid and disk identification	6	3	5	90
Balance not calibrated	7	2	8	112
Wrong oven temperature	5	3	7	105
Calculations	2	2	5	20

7.15 Sampling Frequency

At this point the management has to determine the frequency of control of particular risk, where frequency is the minimum rate that a laboratory action should be sampled for analysis against the particular potential defect.

As shown above, in order to produce a simple quantitative risk assessment, a simple numerical scale as the one reported below can be drawn. In this simple case we assigned the number from 1 to 3 where 1 is the least harmful and 3 the most, and where 1 is the least likely to occur and 3 the most likely. From this the overall risk assessment can be expressed as the product of the two factors, yielding an overall numerical score (1 to 9), as in Table below:

7.15.1 Quantitative Overall Risk Assessment (Simplified)

	Value	Slightly harmful 1	Moderately harmful 2	Extremely harmful 3
Unlikely to occur	1	1	2	3
Likely to occur	2	2	4	6
Very likely to occur	3	3	6	9

In this the final risk 'score' for an extremely harmful contaminant that is unlikely to occur appears identical to that for a slightly harmful contaminant that is very likely to occur. This may not necessarily be a true equality of risk, but is simply the scoring by this very simple risk assessment model.

7.16 Frequency of Sampling/Analysis

At its simplest the translation from risk assessment value to sampling frequency could be a direct relationship, so that in a given period the particular risk could be subjected to the defined analysis the number of times indicated by the numerical risk assessment figure. Using Table above as an example, if the period is assigned as 1 year a trivial risk, it would be sampled once a year, low risk twice in the year, moderate risk 3–4 times, substantial risk every 2 months and intolerable risk every 6 weeks.

This linear relationship in practice may give insufficient weight to the higher risk situations, or may give too much weight to the lower risk situations. It should, however, be feasible to develop a model that uses an appropriate mathematical function to weight the figures to give a statistically sound chance of detecting the faults that is proportionate to the risk.

Table 7.2 Example of possible derivation of sampling frequency (notional)

	Slighly harmful	Moderately harmful	Extremely harmful
Unlikely to occur	Trivial risk: Sample every 3 years	Low risk Sample once a year	Moderate risk 1 Sample every 5 months
Likely to occur	Low risk Sample once a year	Moderate risk 2 Sample every 2 months	Substantial risk Sample weekly
Very likely to occur	Moderate risk 1 Sample every 5 months	Substantial risk Sample weekly	Intolerable risk Sample twice per day

For a truly effective control system, instead of using a linear scale, an exponential scale could be more appropriate.

For instance, a weighting factor, see table below, can be designed to give an exponential relationship that will give a greater weighting for higher risk than for lower risk when compared with the linear example. The numerical value is given as the mathematical exponential constant *e* raised to the power of the value found in the above. Results = e^{x} where x = product of assigned values (1–9) and *e* is the mathematical exponential function (2.72). See table below.

7.16.1 *Quantitative Risk Assessment (Simplified) Incorporating Example Exponential Weighting Function (Notional)*

Value	Slightly harmful 1	Moderately harmful 2	Extremely harmful 3
Unlikely to occur 1	2.7	7.4	20
Likely to occur 2	7.4	55	403
Very likely to occur 3	20	403	8103

To translate these figures into sampling rates, if it were desired to cover the lowest risk at least once in, say, over 3 years' period, dividing all the above by 2.7 (the value for the trivial risk) gives the number of times every 3 years that the risk would need to be sampled, giving the following frequencies in practice (Table 7.2):

7.17 Useful Reading

ISO 31000:2018- Risk management: Principle and guidelines

ISO 31010:2019- Risk management: Risk assessment techniques

ISO Guide 73:2009- Risk management: Vocabulary

ISO 14971:2019 Medical devices: Application of risk management to medical devices

ISO/TR 24971: Medical devices: Guidance on the application of ISo14971

7.18 Example of Risk Assessment for an Analytical Method from sampling Collection to Test Results

	Lab risk assessment for rice and mill humidity analysis									Doc N°.....
			Initial risk assessment				Final risk assessment			
N°	Activity/Process	Risk	Severity	Probability	Risk rating	Control in place	Detection	Severity	Probability	RPN
1	Sample collection	Misidentification of sample	2	2	4	Trained person for receiving samples. SOP for sample identification and reporting (Sample logbook, code and so on)	1	2	2	6
2	Sample collection	Not stored in appropriate space	4	2	8	Sample are stored in designated area. Environmental conditions controlled and recorded every day				
3	Sample preparation	Process parameters: Temperature; weighing errors, reagents	4	2	8	Weighing balance are calibrated and positioned in appropriate location. Volume measurement made with calibrated flask. Reagents are available and used as required. Expire dates are controlled. Personnel well trained for performing the test				
4	Performing test	Sampling homogeneity				Refer to previous similar samples for weight. Refer to Ingamels test	8	5	3	120
5	Performing test	Lid and disk identification				Double control in place	5	3	5	75
6	Performing test	Balance not calibrated				Lab instrument are calibrated from accredited companies for accurate measurement. Label with the date of the next calibration	4	2	8	48
7	Performing test	Wrong Oven temperature				Lab instrument are calibrated from accredited companies for accurate measurement. Equipment log book are available in which all data are recorded on weekly basis	5	3	7	105
8	Preparing test results	Miscalculation data entry	2	2		Final report is controlled by the lab responsible and approved by the Technical manager	5	2	2	20
9	And so on									

Lab risk assessment for rice and mill humidity analysis											
				Initial Risk Assessment (without any controls in place)				Residual Risk Assessment			
Sr #	Area	Activities / Process	Quality Concerns / Risks	Severity Rating	Likelihood	Risk Rating	Existing Controls in Place	Severity Rating	Likelihood	Risk Rating	Further Controls Required
1	Lab Activities	Sample collection	Misidentification of sample/specimen	3	3	9	Sample receiving person from lab verify name & identification Criteria. Furthermore, as sample received in lab it is entered in Sample Log and batch number/code is also alotted otherwise he will return sample to customer for proper identification	3	1	3	
2	Lab Activities	Sample collection	Samples received in lab is not placed in controlled environment	4	2	8	All sample are stored in designated location of lab and environment conditions of lab is recorded on daily basis. Lab Incharge also verified environment record.	4	1	4	
3	Lab Activities	Sample preparation	Inaccurate process parameters (e.g. Temp. Pressure, weighing error), expire reagents used in preparation	4	2	8	Weighing balance are calibrated and placed on flat surface. Furthermore rubber pad under weighing balance to reduce vibrations and it is place in designated position. Volume measurement is done through calibrated flask. Chemist are properly trained on testing activities. Reagents lists are available used in testing activities, expiry of reagents are recorded in this list. Before using reagents Chemist verify its expiry	4	1	4	

4	Lab Activities	Performing Test	Inaccurate test parameters wABC e performing test	4	2	8	Lab STMs are developed and available at the place of use. Active testing parameters are saved in HPLC. Chemist just select the test parameter from equipment for testing. To ensure accuracy of test results lab run standard with sample	4	1	4	
5	Lab Activities	Performing Test	Lab person perform test in lab without considering environment conditions	4	2	8	Lab environment is controlled and its conditions is recorded on daily basis. Lab Incharge also verify evironment record.	4	1	4	
6	Lab Activities	Performing Test	Inaccurate measurements	4	1	4	Lab Instruments are calibrated from accredited lab for accurate measurements	4	1	4	
7	Lab Activities	Chemical Spillage	QC Lab Staff health & work risk, Lab environment condtion and result error may occur	3	4	12	Lab has chemical spill kit available and Emergency eye wash is placed to avoid serious injury.	3	1	3	
8	Lab Activities	Preparing Test Results	Lab Person prepare test report without performing test on specimen	5	2	10	Equipment log books are available in which all data of equipment are recorded on daily basis. Chromatrograph results of actives are attached with final report to ensure the integrity and these chromatrographs have date and time mention on it	4	1	4	
9	Lab Activities	Preparing Test Results	Miscalculation Data entry mistakes wABC e preparing results	5	2	10	Final report is approved by GM Technical/Manager QC after verification of test evidence and reviewed by Chemist. If evidence is not appropriate report will not be approved. All test results from raw data and from equipment data. If it is not match with raw or machine data then Chemist correct typo error	5	1	5	

10	Confidentiality	Lab Information Leakage	StaffofQClableaks confidential information of test results and equipment's	4	2	8	Lab Staff has implemented Confidentiality Policy Furthermore all confidential information is under control condition	4	1	4	
11	Confidentiality	Lab Information Leakage	Personnel other than lab having authorization of lab data (IT etc.) leak confidential data	2	5	10	Lab has signed NDC agreement from all authorized person. In case of any violation legal action can be taken against person	2	2	4	
12	Confidentiality	Lab Information Leakage	Lab confidential data theft in electronic or hard form	2	5	10	PCs are password protected and hard form documents are under lock & key. Authorized person have signed Non-Disclosure (NDC) agreement	2	2	4	
13	Impartiality	Impartiality	Staff of QC Lab should be independent from the production.	4	2	8	Lab Staff has implemented Impartiality Policy. Lab is independent of external or internal pressure and conflicts	4	1	4	
14	Impartiality	Performing Test	Lab Person adulterate the report due to good relation with customer	4	2	8	Test report adultration is not possible in case of active because its result is automatically generate from system. In case of weighing, volume, pH log books are maintained. Furthermore	4	1	4	
15	Impartiality	Ownership	Due to per pressure from company owner; top management interfere to change test results	2	5	10	Lab Authorized persons and management have signed non-disclosure agreement with company in case of violation legal action can be taken against person. ABC has good repute in market so top management/ owner can't force to adultrate results to avoid minimal loss as compare to company repute				

16	Impartiality	Shared Equipment	Due to un-availability of some desire equipment; Lab use production or store department equipment for its activities	3	5	15	ABC lab has dedicated equipment in labs that is not available in other department. Other than this supporting equipments like monitors etc. back up are available in lab	1	2	2
17	Impartiality	Shared facility	Lab facility is used by production or store department for its activties	4	5		Lab facility is dedicated for testing activities only and persons are authorized for this. Unauthorized persons are not allowed in lab	2	2	4

RISK MATRIX

Risk Assessment Matrix Level of Risk	IMPACT				
PROBABILITY	Very Low 1	Low 2	Medium 3	High 4	Very High 5
Very High (5)	5	10	15	20	25
High (4)	4	8	12	16	20
Medium (3)	3	6	9	12	15
Low (2)	2	4	6	8	10
Very Low (1)	1	2	3	4	5

Value	Probability	Definition
5	Very High	At least once in a Day
4	High	At least once in week
3	Medium	At least once in a month
2	Low	At least once in 6 months
1	Very Low	At least once in a year or Rarely Occurs

Value	Impact	Description
1	Very Low	Little/No impact on testing activities or the personnel performing the activity
2	Low	Low Impact due to changes in work environment which may create delays in testing activity
3	Medium	Medium Impact due to changes in work environment (or equipment) which may lead to ambiguous testing results
4	High	High Impact due to personal grievances or pressure on the personnel performing the test which may lead to compromised test results or falsified report ultimately lead to high financial loss > 500000
5	Very High	Very high Impact due to illegal activity / Bribery case / compromise/change in report for personal gains

Index

S. Mannino, *Guidelines for Laboratory Quality Managers*, Integrating Food Science and Engineering Knowledge Into the Food Chain 14,
https://doi.org/10.1007/978-3-031-11724-4